Christian Hambeck

Wake-up Receivers for Wireless Sensor Networks

Christian Hambeck

Wake-up Receivers for Wireless Sensor Networks

Survey, Design and Implementation

Südwestdeutscher Verlag für Hochschulschriften

Impressum/Imprint (nur für Deutschland/only for Germany)
Bibliografische Information der Deutschen Nationalbibliothek: Die Deutsche Nationalbibliothek verzeichnet diese Publikation in der Deutschen Nationalbibliografie; detaillierte bibliografische Daten sind im Internet über http://dnb.d-nb.de abrufbar.
Alle in diesem Buch genannten Marken und Produktnamen unterliegen warenzeichen-, marken- oder patentrechtlichem Schutz bzw. sind Warenzeichen oder eingetragene Warenzeichen der jeweiligen Inhaber. Die Wiedergabe von Marken, Produktnamen, Gebrauchsnamen, Handelsnamen, Warenbezeichnungen u.s.w. in diesem Werk berechtigt auch ohne besondere Kennzeichnung nicht zu der Annahme, dass solche Namen im Sinne der Warenzeichen- und Markenschutzgesetzgebung als frei zu betrachten wären und daher von jedermann benutzt werden dürften.

Coverbild: www.ingimage.com

Verlag: Südwestdeutscher Verlag für Hochschulschriften GmbH & Co. KG
Heinrich-Böcking-Str. 6-8, 66121 Saarbrücken, Deutschland
Telefon +49 681 37 20 271-1, Telefax +49 681 37 20 271-0
Email: info@svh-verlag.de

Approved by: Wien, TU, Diss., 2011

Herstellung in Deutschland:
Schaltungsdienst Lange o.H.G., Berlin
Books on Demand GmbH, Norderstedt
Reha GmbH, Saarbrücken
Amazon Distribution GmbH, Leipzig
ISBN: 978-3-8381-2322-6

Imprint (only for USA, GB)
Bibliographic information published by the Deutsche Nationalbibliothek: The Deutsche Nationalbibliothek lists this publication in the Deutsche Nationalbibliografie; detailed bibliographic data are available in the Internet at http://dnb.d-nb.de.
Any brand names and product names mentioned in this book are subject to trademark, brand or patent protection and are trademarks or registered trademarks of their respective holders. The use of brand names, product names, common names, trade names, product descriptions etc. even without a particular marking in this works is in no way to be construed to mean that such names may be regarded as unrestricted in respect of trademark and brand protection legislation and could thus be used by anyone.

Cover image: www.ingimage.com

Publisher: Südwestdeutscher Verlag für Hochschulschriften GmbH & Co. KG
Heinrich-Böcking-Str. 6-8, 66121 Saarbrücken, Germany
Phone +49 681 37 20 271-1, Fax +49 681 37 20 271-0
Email: info@svh-verlag.de

Printed in the U.S.A.
Printed in the U.K. by (see last page)
ISBN: 978-3-8381-2322-6

Copyright © 2011 by the author and Südwestdeutscher Verlag für Hochschulschriften GmbH & Co. KG and licensors
All rights reserved. Saarbrücken 2011

Preface

The first chapter of this book introduces into the topic of wireless sensor networks. It presents an overview about typical applications and resultant requirements at first. The set of problems regarding wireless networks with very low energy budget and concurrently long lifetime is explained. Possible solution approaches are discussed and the tasks and goals of this work are given.

Chapter two presents state-of-the-art of sensor nodes and their subunits with the help of commercial products and scientific literature. Starting from system level, the fundamental components of a wireless node are considered with the focus on power consumption and energy management. The main part deals with the most important concepts for wake-up receivers from already published literature. Performance advantages and drawbacks are analyzed and discussed, and aspects of technological integration are compared. At the end of this chapter, requirements and challenges for wake-up receiver designs are summarized.

Chapter three explains the proposed concept for the integrated wake-up receiver solution by means of logical argumentation. Beside design process and architecture, the strategy for further power saving is presented in detail. The innovative concept and numerous optimized subsystems allow for a practical full-fledged solution for microcontroller operation, and enable a level of power consumption well below actual state-of-the-art.

Chapter four comprises implementation of the proposed receiver architecture into integrated circuits. After prove of simulation accuracy via measurements on discrete assemblies, the focus of this chapter is put on ultra-low power design of predominantly analog integrated hardware components. The physically given sensitivity boundary for the chosen receiver architecture is evaluated by means of theoretical calculation and compared to simulation results.

Chapter five presents and discusses measurement results of the realized wake-up receiver concept. Detailed comparison of expected results from simulation with measurements and theory allows benchmarking of the implementation from chapter 4. Thereby, all relevant subsystems of the wake-up receiver are characterized and analyzed. Finally, the overall system performance is compared to state-of-the-art and the achieved key parameters and measured values are given.

Chapter six summarizes this work and presents the most important results. An outlook explains possible extensions for the developed wake-up receiver, and sketches its potential with the help of novel and seminal application fields.

Acknowledgements

I would like to thank all those who have contributed to the success of my PhD thesis with their professional or personal support.

My first thanks goes to my doctoral advisor Dr. Dietmar Dietrich for excellent supervision and numerous enlightening discussions with invaluable feedback throughout the course of my work. Furthermore, I would like to thank Dr. Keith Dimond for his expert opinion on the thesis. Many thanks go to Dr. Stefan Mahlknecht and all colleagues from the Institute of Computer Technology at Vienna University of Technology for their continuous help and support.
Special thanks I would like to give to Thomas Herndl from Infineon Austria AG for the outstanding collaboration and his assistance for ASIC fabrication.
Last but not least I would like to thank my family and my friends for their support and motivation during preparation of my thesis.

This work was partly funded by the European Commission within the CHOSeN research project [1].

Table of Contents

1 Introduction — 1
  1.1 Research Activities and Nature of Wireless Sensor Networks — 1
  1.2 Application Fields of Wireless Sensor Networks — 3
  1.3 Network Characteristics and Node Requirements — 8
  1.4 Problems, Challenges, and Objectives — 11

2 Related Work — 15
  2.1 Sensor Node Architecture — 15
    2.1.1 Transceivers — 17
    2.1.2 Application Controllers — 20
    2.1.3 Sensors — 22
    2.1.4 Antennas and Radio Links — 24
  2.2 Node Power Management — 26
    2.2.1 Energy Sources — 28
    2.2.2 Power and Energy Demands — 31
    2.2.3 Medium Access Control Protocols — 34
  2.3 Wake-up Receiver — 38
    2.3.1 Concepts and Performance — 38
    2.3.2 Requirement Analysis — 51

3 Solution Concept — 56
  3.1 Design Process — 56
  3.2 System Architecture — 60
  3.3 Strategies for Power Saving — 65
  3.4 Radio Frequency Frontend — 72
    3.4.1 Envelope Detector — 72

		3.4.2	Noise Considerations	74
	3.5	Baseband Signal Processing		76
		3.5.1	Mixed-Signal Correlation Unit	80
		3.5.2	Digital Correlation	82

4 Implementation   84

4.1	Discrete Measurements			84
4.2	ASIC			86
	4.2.1	Block Diagram and Interfaces		87
	4.2.2	Selected Schematics		89
	4.2.3	Layout		99
	4.2.4	Register Map		101
4.3	System Simulation			104
	4.3.1	Analog Frontend		105
	4.3.2	Mixed-Signal Correlation		106
	4.3.3	Key Parameter Summary		108

5 Measurement Results and Discussion   110

5.1	ASIC Characteristics		110
	5.1.1	Measurement and Test Environment	110
	5.1.2	Analog Frontend	112
	5.1.3	Backend	121
	5.1.4	Wake-up Receiver System	124
5.2	Comparison		128
5.3	Performance Summary		130

6 Conclusion and Outlook   133

6.1	Summary and Conclusion	133
6.2	Enhancements and Future Work	137
6.3	Outlook and Vision	138

Literature   140

Internet References   148

List of Abbreviations   150

1 Introduction

One consequence of modern silicon technologies with low power consumption are wireless sensor networks (WSNs). A WSN is a network of autonomous and distributed sensor devices with wireless communication links and usually self-sufficient energy supply. The main objective of such a network is to collect and process sensor data from environmental conditions and forward it to a superior unit that provides connectivity and interaction to an application service. Typical fields of application are industrial process control, building and home automation, measuring tasks in automotive and aeronautic traffic, or tracking and monitoring in logistic systems [SMZ07]. Usually, a sensor node is equipped at least with a radio transmitter, a radio receiver, a micro controller, an energy source and one or more sensor devices which are assembled to a small footprint device for integration into the environment.

1.1 Research Activities and Nature of Wireless Sensor Networks

WSNs cover very broad fields of research and development topics. Starting from application design and an architecture point of view, hardware design of radio frequency (RF) transceivers, data and signal processing units, supporting sensor devices, energy supplies as well as system integration play an important role. Even more, firmware and protocol design cover a huge range of possibilities for investigation. For a specific application, all of the mentioned fields have to play together in an almost optimum way for good overall performance. As a consequence, many research projects are currently ongoing with versatile aspects and focus of interest to exploit the future potential of WSNs. The first of a few selected ones with relevance to this work is the European Union funded eCubes project [2]. It deals with high density system-in-package (SiP) integration by means of chip stacking for application in harsh environments.

Introduction

One related Austrian funded research project was PAWiS [3]. Its consortium partners were Infineon Austria AG [4] and the wireless group from the Institute of Computer Technology, Vienna University of Technology. The project focused on a power simulation framework for WSNs and on the design of ultra-low power hardware for sensor nodes. Thereby, a power management and power supply unit with support for energy harvesting devices [HHJ+08] as well as a wake-up receiver [SDM09] were developed. Another EU funded research project is CHOSeN [1]. It comprises automotive and aeronautic application scenarios with in-car communication and structure health monitoring for aircrafts. The project target covers hardware design, development of protocols and middleware and the setup of a flexible demonstrator system for the specified applications, whereas design and realization of an ultra-low power wake-up receiver (WuR) are part of the research project.

For illustration, figure 1.1 shows a typical link topology of WSNs. Usually, wireless nodes communicate with one or more gateway nodes that commonly have a wired power connection. The wireless link may be directly to the sink node, or the network establishes a multi-hop connection path via the nodes in the middle that act as routers. Due to high mobility of sensor nodes, possible connection schemes for the wireless nodes range from fully meshed architectures with redundant links to topologies with temporary isolated nodes or node clusters. This brings additional complexity into protocol design for network and routing layers.

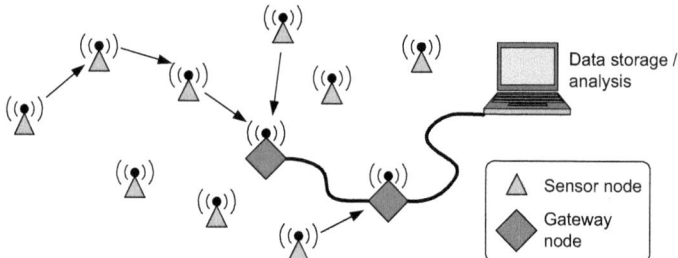

Figure 1.1: Typical topology of a wireless sensor network with single-hop and multi-hop communication links and wired gateway nodes

Since the sensor nodes operate wireless, an energy supply unit has to provide the necessary power. In the simplest way, the node is powered by a cable connection. If a node operates self-sufficient, either a primary battery has to provide energy or energy harvesting and energy buffering techniques have to be used. To extend network lifetime and services to a maximum without battery replacement, it is obvious that power consumption and energy efficiency are one of the most important design parameters for WSNs. They have impact to nearly every module in hardware and software and especially to communication protocol design issues. In many use cases, the energy supply of wireless nodes should last for the whole projected application lifetime and this is often multiple years. So reduction of power consumption, enhancements in efficiency and advanced power management techniques are major goals in research to further extend network lifetime and performance on the one hand, and get a reduction of size, weight and cost of energy storage devices or energy harvesters on the other hand.

This chapter gives an introduction into the topic of WSNs. On the basis of a few selected typical applications in presence and for novel use cases in future, an overview of network and node characteristics in different environments is given. Furthermore, the contribution of this work to the field of WSNs is outlined and the goal of reducing the power consumption for sensor nodes with limited energy budget is presented. This is achieved by an innovative transceiver architecture which promises better network performance tradeoffs when compared to traditional transceivers. New ideas and the approach for future work in power saving are given.

1.2 Application Fields of Wireless Sensor Networks

In the context of wireless sensor networks, the term of ubiquitous sensor networks (USNs) is often mentioned. It describes a more generalized view of sensor networks including wired infrastructure, while this work focuses predominantly on wireless nodes or the wireless part of USNs. The whole field of possible applications for WSNs is very broad and ranges for example from simple passive or active radio frequency identification (RFID) tags for anti-theft systems to large, complex and heterogenous systems in transportation sector with highly dynamic operation requirements. Without a claim for completeness, some typical application systems are introduced to get an insight into the kind of applications to which this work focuses.

Structure health monitoring is one novel application with growing market potential. Maintenance of large buildings such as bridges and vehicles like trains, ships, or airplanes with permanent usage and long projected lifetime is very cost-intensive. As a consequence, supervision can be done by WSNs with a large amount of sensor nodes to determine the optimum time-point for maintenance. Mostly, the sensor nodes are integrated into the structure and cannot be replaced, so their lifetime must be guaranteed over the target's period of operation. Mechanical stress of bridges or skyscrapers is monitored, which may be caused by load conditions, aging or earthquakes [KPC+07]. This goes up to prediction of corrosion effects via monitoring of environmental climatic conditions. The major job of such networks is to collect sensor data and forward it to a central unit for analysis and reporting. In the CHOSeN research project, the partner EADS [5] investigates structure health monitoring systems for aeronautic application.

Introduction

Figure 1.2(a) depicts a large sensor network of up to 5000 nodes that covers the entire aircraft. The used types of sensors are commonly very simple mechanical ones with the advantage of very low power consumption to fulfill the stringent requirements for long network lifetime, size, and especially for weight, in order to minimize additional fuel costs. So called "crack-wires" and torque sensors are bistable mechanical sensors that trigger only once when a certain threshold of expansion or torque is exceeded. An analysis unit then determines the structural part to replace at maintenance, if cumulative defects occur within a certain area. Another application

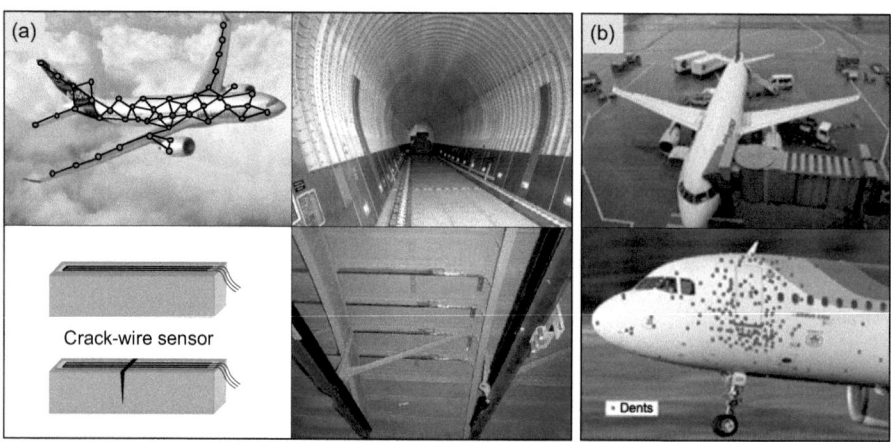

Figure 1.2: Aeronautic structure health monitoring from [HZK+09], [1]: (a) sensor network for detection of mechanical stress and defects within the plane's body via crack-wire sensors and (b), illustration of dents at the door surrounding structure that result from touches with the gangway

illustrated in figure 1.2(b) is a door surrounding stress measurement system. During an aircraft's lifetime, numerous touches and hits to the door environment result from attachment of the gangway and lead to dents and debilitation of the surrounding structure. This is a major problem for the anyway weak point of a door aperture, so a network of deformation and acceleration sensors is used to collect measurement data at high sample rates as soon as the gangway is attached. Because of complex and power hungry analysis of measurement data, the calculation is done by a powered gateway node. For both scenarios described, the energy supply comes from thermoelectric harvester devices that accumulate the electric power generated from the temperature gradient at starts and landings of the aircraft [BKS+09]. This is necessary because of the lack of suitable energy storage devices guaranteeing sufficient power over up to 40 years of operation, which is required for structure integrated nodes.

In the automotive application sector, development tends to a reduction of electric cabling to an absolute necessary minimum. Nowadays, a cable tree of modern cars contains of several kilometers of copper wires. Production, assembly and error diagnosis of cable trees even with different options for various car models are very cost intensive, so manufacturers try to get rid of the heavy and very fault-prone wiring and connectors and replace them with wireless communication links. The scenario in figure 1.3(a) illustrates in-car communication with wireless sensor nodes. Typically, noncritical comfort features such like rain and fog sensors, sensors for air condition, parking assistance, occupation detection, keyless entry or mirror control can be supported by wireless communication links at least. But also parts of safety relevant systems (e.g. a collision warning system [HZK+09]) get more and more assisted by numerous heterogeneous wireless sensor devices that provide additional measurement data of environmental conditions for decision making [6]. Another important application from figure 1.3(b) is a tire

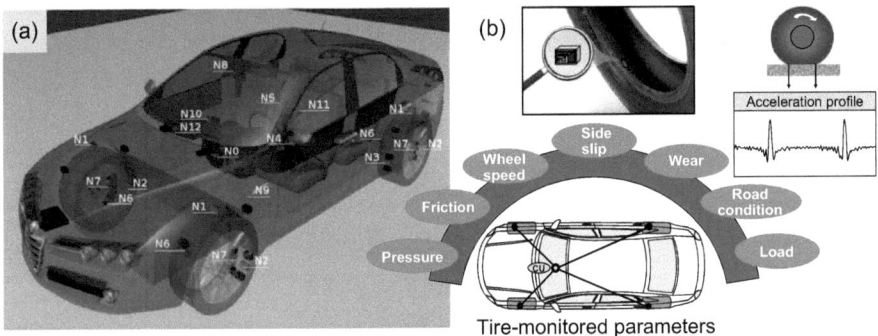

Figure 1.3: Automotive application examples from [HZK+09, HHJ+08], [6]: (a) in-car wireless communication link between central gateway node N12 and sensors for road water (N1), road temperature (N3), external air temperature (N4), fog and rain level (N8), break temperature and pressure (N7, N6), tire temperature and pressure (N2), human audio/video/tactile interfaces (N10, N11), and (b), advanced tire pressure monitoring system (TPMS) for in-tire application

pressure monitoring system (TPMS). In contrast to common state-of-the-art rim-mounted tire pressure sensors, in-tire mounted sensor devices are able to provide much more and accurate measurement data. With the help of pressure, temperature and mainly acceleration sensors, it is possible to calculate the tire footprint and estimate road conditions precisely. This is of major interest mainly for enhancements in breaking systems. Challenges of such systems are the requirements for a volume smaller than $1\,cm^3$, a system weight below 5 grams, the very harsh environment with high temperature range, and reliability to mechanical vibration and shocks of up to $2000\,g$ [HHJ+08]. Since conventional batteries cannot fulfill these requirements, an energy harvester in combination with a storage capacitor provide the power supply for this node. A micro electromechanical system (MEMS) harvester utilizes an electrostatic transduction principle to scavenge energy from mechanical vibration. It is stacked within a molded SiP together with application specific integrated circuits (ASICs) for sensing, radio transceiver and micro controller [2]. Further automotive application examples are inter-car communication for traffic jam detection as well as traffic surveillance and control for flow optimization via an adaptive traffic management system [SMZ07, RSK+07].

Building automation is a further growing field for WSN application. The "smart buildings" may be a skyscrapers, office buildings, or homes that are equipped with various types of sensors and actuators for monitoring and control of different environmental parameters [Far08]. The field of functions usually ranges from support of comfort, security and safety features and goes up to optimization of running-costs. Regardless of a novel application spectrum, wireless interconnection infrastructure is able to replace a large amount of common cable connections. Therefore costs can be saved during construction and particularly in case of rebuilding [Gut04]. Figure 1.4 depicts some examples for intelligent surveillance and control systems [Mah04]. Intelligent power and energy management and electrical load control also in home automation

Figure 1.4: Wireless sensor network examples in building automation

become more and more import. Smart metering and demand-side management prove significant potential for reduction of average power consumption and energy cost [VGGC07]. In future, the influence of green power generation will grow and move from centric towards domestic topologies. To be able to guarantee power quality and control stability within the power grid, it is necessary to get sufficient information and to be able to control demand. Such management systems can be supported by WSNs for environmental measuring tasks at demand-side and control of thermal loads on the one hand. On the other hand, sensor networks may be used for prediction of weather dependent power generation for solar energy or wind energy.

For logistic systems, tracking and parameter monitoring of cargo is an important issue (see figure 1.5). Thus, observation of containers and truck type mountings across the route of transportation can help to minimize delays in delivery, detect misdirection and report alarm conditions by means of sensing [7]. Environmental measurement data such as temperature, humidity, acceleration profiles and position can be logged, and may be of interest for the transport insurance [MM07]. Further on, advanced systems with distributed sensors can provide intrusion detection and theft detection service for containers. Typical nodes are equipped with global positioning system (GPS) and a GSM/GPRS or UMTS modem. They must guarantee service-free operating lifetimes of at least 5 years to cover a container maintenance interval also with the power hungry cellular network connection.

Figure 1.5: WSNs for container monitoring and tracking system: container port in Hamburg, and network topology for in-container observation system with external wireless communication link

Further applications for WSNs are in the field of environmental technology. This may include systems for pollution detection, glacier monitoring, detection of temperature distribution, forest fire recognition, systems for prediction of volcanic eruption or agricultural support systems that monitor rainfall and sunshine duration.

In medical health care, sensor nodes are able to monitor activities of needy persons or vital signs such as heart beat, body temperature, blood pressure, or arterial oxygen saturation [VWS+06]. Thus, safety in home care can be increased.

Military use cases for WSNs are surveillance of battlefields and mine-fields, sniper localization or vehicle tracking [RM04].

1.3 Network Characteristics and Node Requirements

As introduced in the previous section, the whole range of applications for WSNs is heterogeneous, each with very different nature and also more and more novel developments are emerging in future. In contrast, the requirements for a dedicated solution are specific and have to meet the focus of interest in an almost optimum way. The following list points out major network characteristics and node requirements with main respect to power consumption. Their consequences for WSNs are outlined and figure 1.6 illustrate primary interdependencies.

- Network topology: In WSNs, schemes for wireless communication links of sensor nodes range from fully meshed topology with redundant connections up to multi-hop links or temporary isolated network islands. Nodes can join or leave the network, so all connections are time-variant. Since WSNs are ad-hoc networks, one challenge are efficient routing mechanisms and protocols to cope with all possible application scenarios, and simultaneously meet the application demands for fast network (re)establishment at low traffic and protocol overhead. Consequently, the buildup of robust communication links has main impact to overall energy consumption because of the extensive transceiver activity.

- Long network lifetime: Since most of the network nodes are battery powered, a long lifetime is one main goal to extend service duration and reduce the size of batteries or energy harvesters. A sensor network has various operating states with a high dynamic range in power demand profile. Typical modes of operation are high performance calculation, RF-transmission/reception, sensor/data acquisition, idle listening, and deep sleep mode. In each power state, requirements regarding node architecture, network protocols, timing and mainly power management may be different. One solution can be efficient for a certain state, but inefficient for others. So adaptive and configurable structures in hardware and software can help to identify and eliminate bottlenecks that degrade performance. Normally, low duty cycle operation and low average data traffic allow for entering low power sleep modes where most of the power consuming parts are switched off. Nevertheless, the requirements for lowering energy/power consumption lead to strong energy management concepts in hardware design and also in protocol design to balance the node's energy reserve, and thus achieve a maximum lifetime for the complete network service.
Beside energy limited lifetime, reliability plays an important role especially for automotive and aeronautic applications. Sensor networks are often exposed to very harsh environments. Temperature ranges from -40 °C to +125 °C in automotive specification or

even more in aeronautic scenarios, high mechanical vibration and shock amplitudes, and cosmic radiation are challenging for acceleration sensors, system packaging and even for energy storage and energy harvester devices. Also aging effects of oscillators, resonators and batteries at corner conditions have to be considered for reliable long-term operation of up to decades.

- Wireless communication performance: In order to maximize the radio link distance at limited transmit power, high receive sensitivity and interference blocking capability are beneficial. Energy can be saved via reduction of transmit power or via keeping the hop-count as low as possible. Beside link distance and high data rates for fast data exchange and short occupancy of radio channels, one important performance parameter for the sensor network is responsiveness. This specifies how fast a node can react on incoming events or data packets and process or forward them. The advantage of low end-to-end delay for communication and therefore realtime capability implies low per-hop delay and low packet latency below a few milliseconds. This can easily be guaranteed by a method of operation where the receiver is listening permanently to the radio channel, but with the disadvantage of high power consumption.

- Flexibility: Because of versatile hardware and firmware needs, (re)configuration options are necessary to provide good performance balance between competitive network or system parameters such as node responsiveness versus power consumption during all modes of operation. The transceiver may be capable of covering multiple frequency bands and/or support different options for RF-modulation schemes and data rates. This may be configured in tradeoff to receive sensitivity. To ensure an optimum ratio of performance versus energy consumption, a flexible and dynamically configurable power management unit can be beneficial. Finally, the possibility of a wireless firmware upgrade for systems with long operation periods can be used to handle upcoming or changed application requirements. At the end, extensive options for (re)configuration allow a flexible compromise for the targeted optimization criteria even in various operating modes and thus improves guaranteed future.

- High system integration density: A sensor node consists of many different assemblies such as radio transceiver, micro controller, sensors, antenna, some discrete components and an energy storage or harvester device. These all have to be packaged with constraints concerning weight and form factor, whereas the package must be designed to withstand the environmental conditions. While some ASIC parts may be integrated into a system-on-chip (SoC) or a SiP together with discrete components, bulky elements like the antenna or the energy source may be part of the package. Due to the usually high quantity of

Introduction

sensor nodes, cost issues of packaging and assembling have significant importance, so developments tend to solutions with potential for high integration density [2].

- Low cost: High volume mass production and large scale integration of most WSN components enable the use of sensor nodes in cost-sensitive applications and also allow for novel use cases. Because overall costs are often a driving concern for enhanced market chance, the balance of performance versus cost influences node design significantly. For example, the mentioned application of tire pressure monitoring with energy harvester is estimated to have an economic turnover of some hundred million pieces to return research and development expenses [2].

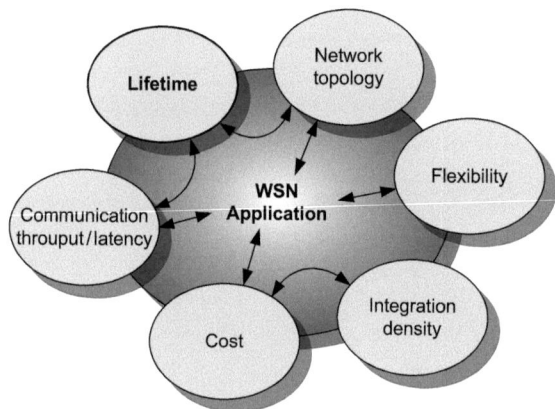

Figure 1.6: Primary requirements and most important interdependencies for WSN applications with focus on node power consumption and thus lifetime

Other design criteria and aspects such as security and safety issues have minor impact to energy consumption and are omitted for simplicity. One can see from figure 1.6 that many of the characteristics and requirements listed above are dependent and interact with each other. So it is not possible to meet all of them at the same time to build the ideal wireless sensor network. The designer always has to make a compromise between tradeoffs. As a consequence, the optimum architecture and characteristics for best network performance are highly dependent to the specific application demands and may also vary within different network states. Configuration options for the sensor node are one possibility to extend the range of acceptable operating conditions to cover the focused application in an almost optimum way. Since most of the energy is consumed typically by wireless communication (see table 2.5, [Mah04]), the goal of this work is to enhance both wireless communication performance and power efficiency of sensor nodes to further extend lifetime of the whole network service. This is achieved via support of an extra hardware option to close the performance gap that results from the common tradeoff between high-performance active mode and low-power idle mode.

1.4 Problems, Challenges, and Objectives

Normally WSNs are powered by batteries or some type of energy harvesters. As a consequence, the available amount of energy is limited and implies one major restriction in node performance. In opposite to permanent supplied devices, this circumstance influences the whole architecture and design of hardware, network protocols, firmware, and application in a crucial way. The objective of this work is to extend the node's lifetime via reduction of power and energy demand without degrading its communication performance.

In opposite to broadband data communication radio links, a sensor network link covers low average amount of data traffic. Furthermore, sensor measurements are typically done periodically in long time intervals. So most of the time, the whole sensor node is inactive and remains in a low power sleep mode where its main parts are powered down completely and power consumption drops down to typically a few micro watts [Mah04, MB04]. Nevertheless, the receiver has to listen for incoming data packets all the time and therefore it has to stay powered and consumes at least around 20 mW (see table 2.1). This usually dominates overall energy consumption by orders of magnitude since energy demand of duty-cycled digital processing is mostly negligible. The traditional solution to cope with this problem are scheduling medium access control (MAC) protocols. They periodically cycle active and shut down mode of the receiver and so communication takes place in specific time slots. In case of figure 1.7 for instance, the receiver is shut down 90 % of the time and active for the remaining 10 %. The average power consumption then is reduced by a factor of 10 when neglecting standby power. But the consequence is that the whole network or at least network clusters have to be synchronized to guarantee, that communication is ensured during the same time slot when all affected nodes are active. Furthermore, latency of packet delivery increases because data transmission has to be delayed until the next time slot. This gets even worse, if the duty cycle is reduced to save more energy. The approach of scheduling MAC protocols has several limitations and disadvantages concerning responsiveness, synchronization, and complexity and is explained in more detail in section 2.2.3.

As Jan Rabaey proposes in [Rab09], the idea to overcome the conflict of power consumption versus latency is an additional receiver, called "wake-up receiver (WuR)". It is a dedicated low-power receiver with the tasks of monitoring the radio channel and generating a wake-up event to notify the sensor node that an incoming data packet is going to arrive. Because the WuR has

Introduction

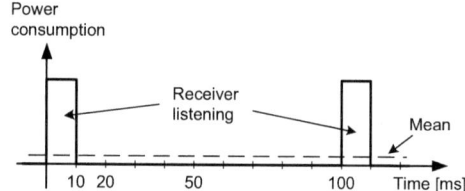

Figure 1.7: Power consumption of main receiver at 10 % on-off duty cycle

very specialized functionality, it can be highly optimized for low power operation. Therefore it can stay powered all the time and ensures high responsiveness, but without dominating the node's energy consumption. Its own-power consumption must be lower by orders of magnitude when compared to state-of-the-art main transceivers and should be lower than 10 µW [MB04]. Figure 1.8 illustrates the operating principle of a WuR based sensor node architecture: Compared to traditional designs, the block diagram of the sensor node is supplemented with the WuR and an antenna switch or power splitter for appropriate allocation of RF power. While the main receiver is put in power down mode during state $S1$ of the timing diagram, the WuR is active and waits for incoming data. Once the WuR receives a wake-up call from the sender, it checks its address information in the wake-up preamble of the sender's message. If an address match is detected, the WuR asserts an interrupt request (IRQ) to the application controller and the main receiver is activated. The subsequent data communication phase is conventional whereat states $S2$ and $S3$ demonstrate data reception and transmission. Finally, the main receiver is set to sleep mode again after communication, and the WuR stays in active listening mode and waits for the next arriving wake-up call. Thus, the node returns to power state $S1$ again.

To the best of the author's knowledge, there is no commercial product of a wake-up receiver for frequency ranges above 100 MHz available up to now that provides high receive sensitivity at a power consumption of a few microwatts. In scientific literature, there exist some implementations (see section 2.3), but a full-fledged companion chip solution for common wireless sensor nodes has not been reported yet. Important work was done at Berkeley Wireless Research Center, University of California. Their WuR realizations [PGR07, PRG09] reached power consumption levels down to 52 µW and a receive sensitivity of -72 dBm. Marco Spinola-Durante, a former PhD student at the Institute of Computer Technology, Vienna University of Technology developed a WuR prototype with currently lowest power consumption in literature of just 12.5 µW at -57 dBm receive sensitivity [SDM09, SD09].

So the goal of this work ist to investigate and realize an ASIC implementation of a WuR concept with power consumption clearly below the threshold of 5 µW and further push the receive

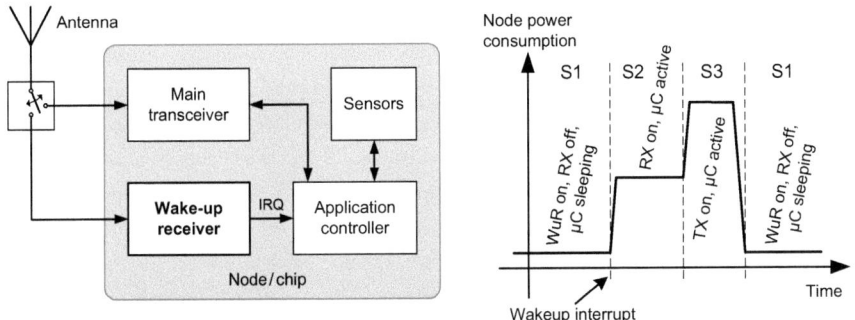

Figure 1.8: Sensor node architecture based on an additional wake-up receiver (WuR): block diagram and corresponding timing diagram for a wake-up scenario
RX: receiver, TX: transmitter, µC: microcontroller

sensitivity of existing solutions to regions where the WuR is applicable for practical scenarios under real world conditions. In order to allow adequate realtime capability, the WuR's maximum latency for wake-up event detection should be below 10 ms. Since the main task of a WuR is only to generate an interrupt request when the device is addressed, it can be designed and optimized for this very specific duty with respect to ultra-low power consumption. The focus of this work is put on analysis, theoretical evaluation of WuR architecture, design of the analog radio frontend, the mixed-signal part, and a baseband signal processing unit. Communication protocol design issues for system integration and optional digital signal processing are addressed, but not discussed in depth in this book. The intention is to implement an ultra-low power consuming solution of high sensitivity and low cost with clearly improved characteristics when compared to state-of-the-art designs that has the potential for almost full ASIC integration into 130 nm complementary metal-oxide semiconductor (CMOS) technology. This WuR design should contain components and interfaces for stand-alone operation and allow a cost-effective companion-chip solution or may be integrated into conventional main transceivers in future.

One major challenge is ultra-low power circuit design. This includes strong power management concepts on one side and many implementation considerations for the specific subunits on the other side to avoid any bottleneck regarding power consumption. Techniques therefore are low voltage design, analog and RF design with logic transistors, weak-inversion CMOS design for digital and analog circuits and mixed signal (pre)processing. The goal is to exploit the semiconductor technology's capabilities as good as possible and investigate ultra-low power circuits close to the technology's limitations. Nevertheless, the outcome must have enough margin to fulfill the requirements for a robust implementation when considering corners of the CMOS process variations and a temperature range of -40 °C to +125 °C from automotive specification. Effects to be taken into account are drain and gate leakage currents of metal-oxide semiconductor (MOS)-transistors and MOS-capacitors especially in analog circuits with high impedance, layout specific parasitic capacitance and resistance, and limitation of chip area for large resistors with inherent large stray capacitance. A further challenge is the aim for a mostly full integrated concept/solution with a minimum number of off-chip components. The implementation should be done in an almost optimum way to get close to the theoretical and physical boundaries of the receiver's architecture. Finally there is always the risk of failure in ASIC design and fabrication with the consequence of time- and cost-intensive redesign and fabrication. The goal is to achieve a "first time right" design.

At the end, nothing is for free, but the goal of this work is to provide a practical solution concept that bridges the performance gap of conventional WSN systems regarding the low power versus low latency tradeoff in order to be able to cover an extended range of application requirements via additional hardware and configuration options.

2 Related Work

A sensor network node is a complex system that consists of many heterogeneous subsystems such as radio transceiver, microcontroller, power supply and management unit as well as sensor interfaces. Its firmware has to handle tasks for measurement and application and it has to cover all layers of the data communication stack. Beside collecting and processing sensor data, a node can also be used for routing, bridging and packet forwarding to extend the radio link range. Since the node operates wireless, its amount of energy is limited which influences the available network service duration in a crucial way. As consequence, additional power and energy management strategies within all layers, starting from hardware equipment up to communication protocol stack are necessary in opposite to mains powered solutions. Accordingly, there has to be a compromise between performance and lifetime of a sensor node for most application scenarios. So careful design and operation with respect to the node's energy reserve has to be considered in nearly all tasks and subsystems.

In this chapter, state-of-the-art literature and products of WSN node components are discussed first. One part deals with strategies and concepts for power and energy management, and the emphasis is put on the core topic of wake-up receiver (WuR) concepts and architectures for WSNs.

2.1 Sensor Node Architecture

A generic architecture of a typical sensor network node is presented in figure 2.1. While the surrounding box covers the core of semiconductor and firmware components, the remaining external devices are at least an antenna for wireless communication, some type of energy source for node supply, an oscillator crystal for generation of the time base, and sensors or actuators for interaction with the environment. The semiconductor core consists at minimum of a wireless

Related Work

transceiver, sensor interfaces and a microcontroller that handles low level node control, communication protocol stack and application firmware. Due to the self-sufficient node architecture, special care is taken on power management. Beneficial therefore is reconfigurable supporting logic that unloads the microcontroller from simple or time critical tasks as well as sophisticated power- and energy management hardware support that saves energy via efficient interplay of harvesting device and energy storage device.

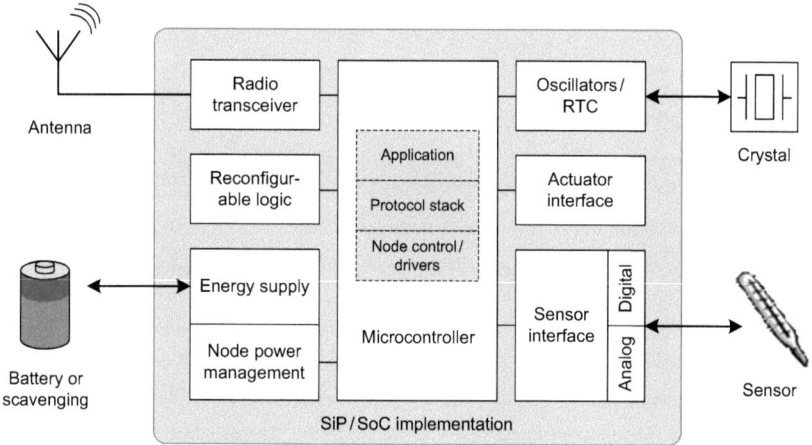

Figure 2.1: Generic sensor node architecture with potential for system-in-package or system-on-chip integration
RTC: realtime clock

Since applications such as building automation or tire pressure monitoring system are cost sensitive and demand a very high number of pieces beyond some millions [KB10], high system integration density is essential. An ideal solution would be a monolithic system-on-chip (SoC). This requires advanced semiconductor technology options that allow for efficient integration of RF circuits, analog components, digital components, as well as memory onto one common chip. One intermediate step towards this way are system-in-package (SiP) solutions such like the application example of TPMS from figure 2.2. The photograph depicts a complete and full functional TPMS sensor node for in-tire assembly within a package cube of 1 cm edge length. Innovative chip stacking and connection technologies help to further reduce size, weight and cost [2].

Figure 2.2: Package integration of tire pressure monitoring system (TPMS) sensor node from [FDH+10], [2]: The pressure sensor and the bulk acoustic wave (BAW) resonator chip are mounted on top of the transceiver ASIC which itself again is assembled on top of the MCU ASIC. This silicon stack is bonded to a printed circuit board and placed inside a molded plastic package of 1 cm edge length. The package includes multi-layer chip capacitors as energy buffer, primary batteries for energy supply and an option to replace the batteries with a MEMS device for energy harvesting.

2.1.1 Transceivers

The task of transceivers is to establish wireless communication links of high quality and reliability between network nodes. For that reason, table 2.1 lists the most important operating characteristics and typical performance values of low-power transceivers, receivers and a transmitter. These parameters are RF modulation capabilities, coverage of frequency bands, available range of data rates, receive sensitivity at a given data rate, startup time, and power consumption. Many state-of-the-art transceivers have very good communication performance, but the main disadvantage is their high power consumption in the range of many milliwatts. The usual approach ist to duty-cycle the receiver to save power. The limits therefore are acceptable communication latency for long power-down intervals and minimum on-time for powering-up the transceiver. At startup, frequency of quartz oscillators and phased locked loops (PLLs) have to stabilize first until the transceiver can be put into receive or transmit mode. This typically takes around one millisecond (see table 2.1) because of the very high Q-factors of oscillating crystals. In opposite to the conventional superheterodyne receiver principle, advanced transceiver architectures and implementations offer new possibilities for integration and power saving.

Promising progress in research has been made with bulk acoustic wave (BAW) based transceiver architectures from figure 2.3 and other implementations [CMP+06]. A BAW resonator is very small compared to traditional surface acoustic wave (SAW) resonators and multiple BAWs can

Related Work

Device	TDA5240 [4]	TDA5150 [4]	ATA542X [8]	BAW based [FDH+10]	BAW based [OCR05]
Type	RX	TX	TX/RX	TX/RX	TX/RX
Main feature	high sensitivity	low power	high sensitiv.	fast startup	low power
Transmit power [dBm]	–	-10 – 10	0 – 10	1.5	-4.0
Frequency band [MHz]	315/433/ 868/915	315/433/ 868/915	315/433/ 868/915	2100/2450	1900
Sensitivity @ 2 kbit/s [dBm]	-119	–	-117	-104	-104.5
Data rate [kbit/s]	0.5 – 112	0.5 – 100	1.0 – 20	50	5
Modulation scheme	ASK/FSK	ASK/FSK	ASK/FSK	FSK	OOK
Supply voltage [V]	3.0 – 5.5	1.9 – 3.6	2.4 – 6.6	2.9 – 3.6	1.0
Supply current [mA]	10.5	9.0	10.5	8.0	0.4
Startup time [μs]	455	1000	1390	2.0	–
Temperature range [°C]	-40 – 105	-40 – 85	-40 – 85	-40 – 125	–

Table 2.1: Overview of characteristics of selected low-power transceivers for WSN
RX: receiver, TX: transmitter, ASK: amplitude shift keying, FSK: frequency shift keying, OOK: on-off keying

be integrated into one small chip [FDH+09]. The significantly higher Q-factor allows for better frequency selectivity and lower loss in filter application. The startup time of BAW based oscillators in the gigahertz range is much lower compared to quartz oscillators. Thus, the power-up time of transceivers can be reduced significantly to only a few microseconds (compare with table 2.1). This saves power even for short power-on intervals of oscillators [DPF+10], but with the disadvantages of high initial frequency tolerance and temperature drift. The transceiver architecture in figure 2.3 from [FDH+10] employs four BAWs resonators on a single die. Two of them are used for pre-selection of frequency band within the low noise amplifier (LNA) stage in the differential designed receive path. A third one defines frequency in the local oscillator (LO) circuit to supply the image rejection mixer that converts the receive signal to intermediate frequency (IF) domain. After IF filtering, a limiting amplifier transforms the signal into digital domain where demodulation, baseband signal processing, and clock and data recovery are done. In transmit path, the BAW based oscillator for generation of RF carrier frequency is directly modulated with transmit data via control of the BAW bias voltage. The frequency modulated transmit signal is then amplified with a power amplifier (PA) an fed to the antenna through a matching network with receive/transmit switch. To compensate absolute accuracy and temperature drift of BAW frequency, a temperature sensor is used together with tunable capacitors in parallel to all resonators for frequency calibration. An alternative possibility for frequency determination and filtering is the application of a MEMS instead of BAW resonators. MEMS based transceiver architectures such as [OCL+04] have a similar benefit of fast oscillator stabilization time. A low power frequency shift keying (FSK) transceiver implementation with passive RX frontend and excellent noise performance is presented in [CBM+06]. Its supply

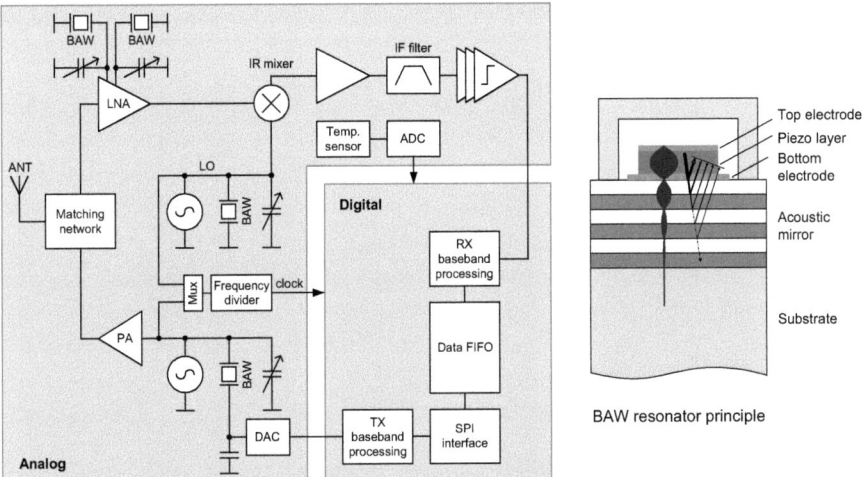

Figure 2.3: Transceiver architecture based on bulk acoustic wave (BAW) resonators, and construction of a BAW filter from [FDH+09, FDH+10], [2]: The BAW resonators are used for direct generation of RF-carrier and LO frequency as well as for filtering purpose within the LNA. They are tuned via variable load capacitance and DC bias voltage in a range of up to 10 MHz for channel selection, modulation and for compensation of tolerance and temperature drift.
The structure in the right-hand drawing illustrates function of a BAW resonator via the vertical acoustic standing wave generated by the piezoelectric layer and reflected by mirror layers.
LNA: low noise amplifier, PA: power amplifier, IF: intermediate frequency, IR: image rejection

voltage is just 400 mV and may be provided by solar cells.

Another trend in transceiver development is to put more and more intelligence into the devices. These "smart" transceivers include support for offloading simple tasks from the microcontroller. This allows for extended MCU sleep periods, saves energy and puts out time-critical constraints from firmware. Those tasks are for example switching between modulation schemes and parameters, efficient handling of power management issues, wake-up condition detection, support for MAC protocol processing and data encryption. The approach is to provide higher level functionality and abstraction via command based interfaces directly to the MCU instead of register map access to simplify programming. Initialization tasks and most of the configuration can be handled by transceiver internally for predefined standards. The transceiver is controlled either by means of intelligent state machines [MDF+10] or a simple central processing units (CPUs) with supporting peripherals [SBK+06]. Texas Instruments [9] offers for example CC2520, a ZigBee [10] transceiver for IEEE 802.15.4 systems. A full featured MCU-transceiver combination is ATmega128RFA1 from Atmel [8]. Amongst others, research in the range of this topic is ongoing within the CHOSeN research project [1].

The parameters from table 2.1 show that all receivers have high sensitivity, but power consumption is in the range of some 10 to 30 mW. They use simple modulation schemes such as amplitude shift keying (ASK), on-off keying (OOK) and FSK, or some variants and combinations of them. Higher modulation formats like pulse amplitude modulation (PAM), pulse position modulation (PPM), orthogonal frequency division mulitplex (OFDM), or code division multiple access (CDMA) are not required, since there is no need for high efficiency in terms of frequency band utilization because of the low data rates in WSN. In addition, the signal processing overhead for complex modulation schemes would further increase power demand dramatically. Beside the transceivers listed in table 2.1, the CC series from Texas Instruments [9] for frequency bands up to 2.45 GHz is one more device family with high performance that is widely used in industry.

If only power consumption is considered for the transmitter, and other aspects are omitted, it makes no sense to reduce transmit power below a certain level. From data sheets of TDA5150 [4] or CC1100E [9] the values for current consumption indicate clearly that power demand of RF-circuitry for generation of carrier frequency and RF-modulation dominates more or less overall consumption, if the transmit power is below 5 dBm for state-of-the-art products up to 1 GHz. Hence lower transmit power would shorten radio link distance with negligible saving of energy. Nevertheless, reduction of RF-power may be useful for some other kind of reason like aware limitation of transmission coverage or minimization of network interference.

2.1.2 Application Controllers

Since demand for processing power and memory space varies in a large range and depend highly on requirements from application, a broad range of application controllers is commonly used. Typical features of microcontrollers with modern architecture and their embedded peripherals are presented in figure 2.4 and table 2.2. Beside pervasive options for bus width of reduced instruction set computer (RISC) CPUs, memory types and processing speed, the assortment of peripheral devices has grown rapidly in the past few years. Numerous types of parallel and serial interfaces support both low-power and high-bandwidth data transfer to external components for communication and storage. Numerous embedded analog extensions for data conversion and power supply allow for high system integration density at low cost even if not all peripherals are utilized for a certain application, because most of the silicon area is normally occupied by memory anyway. In most cases, overall energy consumption of a sensor node is dominated by power demand in sleep mode or idle mode of CPU whereat only selected MCU peripherals, voltage regulators, voltage monitoring circuits and oscillators are active. Then efficiency of computational power is of minor importance, and the focus has to be put on power management. Efficient clock management and power gating for all devices is mandatory. It is gained through fast and direct register access of all embedded peripherals. Offloading CPU via support of configurable auxiliary components enhances system performance on the one hand and can help to minimize power consumption on the other hand. Direct memory access (DMA) controller, timer/counter or accelerators for encryption are classical examples therefor.

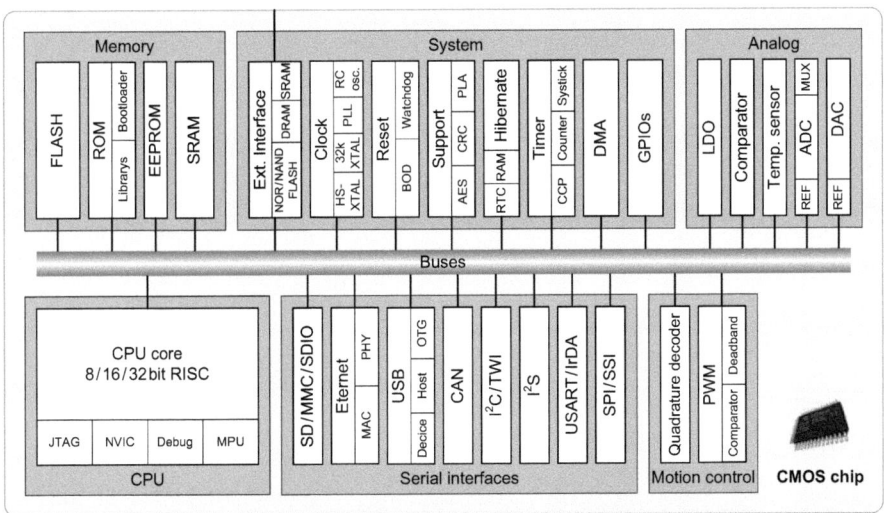

Figure 2.4: Microcontroller architecture with comprehensive peripheral features: The bock diagram illustrates possible functionality and capabilities of modern microcontrollers from [4, 8, 9, 11, 12, 13, 14]. Powerful interfaces and controllers like DMA, DRAM, ethernet MAC or USB host are embedded only in 32 bit microcontrollers, meanwhile smaller devices focus on ultra-low power consumption especially in standby modes. Not all features are available in a dedicated product at the same time, but manufacturer offer various combinations in different product series for nearly all application demands.
XTAL: crystal, RC: resistor-capacitor, PLL: phased locked loop, BOD: brownout detection, AES: advanced encryption standard, CRC: cyclic redundancy check, PLA: programmable logic array, RTC: realtime clock, CCP: capture/compare/PWM, GPIO: general purpose input-output, LDO: low drop output voltage regulator, REF: reference, MUX: multiplexer, NVIC: nested vector interrupt controller, MPU: memory protection unit, I^2C: inter-integrated circuit, TWI: two-wire interface, I^2S: integrated interchip sound, IrDA: Infrared Data Association, SPI: serial seripheral interface, SSI: synchronous serial interface, PWM: pulse width modulation

Related Work

Table 2.2 introduces key parameters of popular WSN microcontrollers for comparison. The range reaches from low-cost 8 bit MCUs up to a high performance controller with 32 bit Cortex-M3 core from ARM [15] that includes a DMA controller, a USB and a memory card interface. If efficiency of computing power is compared, the MSP430 has best performance, but if buswidth and acceleration hardware are considered too for practical use, the 32 bit ARM core is gaining. As stated in [Mah04], the more important impact for overall energy saving in low-power applications has current consumption in standby modes. While all listed MCUs have low consumption in power-down mode at room temperature, current demand in power-save mode with active RTC differs significantly.

Combinations of microcontroller and transceiver to a SoC with the advantage for minimum off-chip component count are CC430Fxxx and ATmega128RFA from Texas Instruments and Atmel.

Device	ATmega1284PA [8]	MSP430F5438A [9]	PIC24F16KA102 [11]	AT91SAM3S4C [8]
Architecture	8 bit RISC	16 bit RISC	16 bit RISC	32 bit RISC
Clock freq. [MHz]	20	25	32	64
Speed [MIPS]	20	25	16	80
FLASH [Kbyte]	128	256	16	256
SRAM [Kbyte]	16	16	1.5	48
EEPROM [Kbyte]	4	–	0.5	–
Main peripherals	3 Timers/WDT RTC 10 bit ADC 6 PWM UART/SPI/I^{2}C	3 Timer/WDT RTC 12 bit ADC 3 channel DMA UART/SPI/I^{2}C	3 Timer/WDT RTC 10 bit ADC PWM/Comp. UART/SPI/I^{2}C	Timer/WDT/RTC SD/I^{2}S/USB 12 bit ADC/DAC 22 channel DMA UART/SPI/I^{2}C
Oscillators	XTAL/RC/32kHz	XTAL/RC/32kHz	XTAL/RC/32kHz	XTAL/RC/32kHz
Supply voltage [V]	1.8 – 5.5	1.8 – 3.6	1.8 – 3.6	1.8/3.3
Current efficiency [µA/MIPS]	400	230	363	450
Power down mode	0.1 µA	0.1 µA	0.025 µA	–
Power save mode (with RTC)	0.6 µA	1.2 µA	0.67 µA	1.78 µA
Temperature [°C]	-40 – 85	-40 – 85	-40 – 85	-40 – 85

Table 2.2: Overview of characteristics of selected low power microcontrollers for WSN
MIPS: million instructions per second

2.1.3 Sensors

The various types of sensors are as heterogeneous as WSNs are. They range from very simple devices with either open or closed electrical contacts up to sensors with much cost in demand of time, energy, and computing power such as GPS for localization. The important point is a balanced tradeoff between low power consumption and sufficient quality of application service and operability. This normally leads to a preferably configurable compromise between performance and power consumption in many cases. Without claim for completeness, the most common types of sensors for WSNs and their characteristics are outlined in the following list.

- Mechanical sensors: Many simple monitoring duties can be covered by mechanical sensors that provide a switching electrical contact. Some examples are crack-wires (see figure 1.2), bistable torque sensors that trigger when exceeding a certain threshold, sensors for intrusion detection with switching contact, or vibration sensors that short contacts during motion via moving mechanical parts. Since checking an electric contact can be done directly by a microcontroller, these types of sensors require very low power consumption if data is sampled infrequently.

- Tracking sensors: Popular techniques for tracking are global positioning system (GPS) satellite receivers, radar sensors for distance and speed control, vision systems with camera, and acoustic systems based on microphones. In opposite to mechanical sensors, the named ones for tracking systems are very cost intensive regarding energy consumption and acquisition time. So turning on the sensor only for a single measurement requires a comparatively high amount of energy. Additionally, there is the necessity for complex processing of raw measurement data to extract the expected parameters from sensor data stream. This yields to a restrained use to meet the energy consumption requirements.

- Optical sensors: Usually low-cost photodiodes and light dependent resistors are used for measuring ambient light intensity. In smoke detectors and gas sensors, the conventional edge-emitting laser diodes are replaced with low power vertical-cavity surface-emitting lasers, that have comparatively low threshold currents down to 0.4 mA such as ULM850-PM from U-L-M Photonics [16]. Together with analog-to-digital conversion at low to moderate sample rates, such kind of sensors can be operated for years also with small button cell batteries.

- Environmental sensors: Humidity sensors, proximity detection devices via pyroelectric principle, semiconductor pressure sensors, and temperature sensors are typical for monitoring ambient conditions with low power consumption. Accelerometers are mostly realized in micro electromechanical system (MEMS) technology. In contrast, they require sample rates of up to a few kilohertz and also signal analysis for each output channel. To overcome this power-wasting fact, manufacturer offer devices with built-in intelligence that can trigger an external interrupt only when exceeding a configurable acceleration threshold. For subsequent evaluation, the previous data sequence is buffered in memory (see ADXL312 sensor from Analog Devices [13]).

- Sensors for metering: For current and power metering tasks, hall sensors are often used with their advantage of galvanic isolation, low loss, and tolerant overload characteristics. If distortion of current waveform in mains supply or active and reactive power are calculated, high sample rates up to kilohertz are necessary to figure out phase shift and harmonics with high accuracy. For metering of mechanical distances and motion, there exist various principles whose applications depend mainly on the environmental conditions. Often used principles for measurement of linear displacement are transducers with optical, ultrasonic, capacitive or inductive concepts. Rotary encoders are usually realized via optical or capacitive sensing principles, whereas power consumption can be very different because of the always-on light source.

There are a lot more types of sensor devices and techniques for measuring physical quantities than presented above, but not all of them are suitable for WSN nodes because of limited resources. At least energy restrictions are often the main reason for changing from a fully wireless system concept to a partly wired architecture, whereat selected nodes with power hungry sensors or data processing units are supplied by mains.

2.1.4 Antennas and Radio Links

Radio link quality is important for WSNs because all nodes share the same medium together with other systems that may cause interference. Superheterodyne reception, normally used in main receivers, has good interference blocking capability in contrast to more simple principles of low-power receivers. Nevertheless, the main limiting factor is usually path loss for radio link channels especially in indoor environment with walls of armored concrete. Since wireless nodes are often integrated into environment and infrastructure, the influence of conductive objects in near-field ($\approx$ wavelength $\times$ antenna gain) has to be considered. For instance, a metal case may shift the antenna's resonance frequency out of band and may degrade radio link performance dramatically. Anyway, objects in radio channel lead to effects of reflection, scattering, diffraction, depolarization and absorption also in antenna far-field, and thus results in additional overall path loss. These effects are frequency dependent, lead to interference and therefore fading of receive signal strength of typically more than 20 dB even for indoor scenarios [OCB+10]. Equation 2.1 describes free-space path loss for isotopic antennas, where d is the distance and

λ the wavelength of the radio link.

$$L = 20 log_{10} \left(\frac{4\pi d}{\lambda} \right) \quad (2.1)$$

The frequency dependent characteristic of path loss is shown clearly even for free-space condition. Table 2.3 illustrates this effect for selected link distances at different industrial scientific and medical (ISM) frequency bands. Lower frequencies are preferred, but with the disadvantage of increased antenna size and weight. The large usable bandwidth of the 2.45 GHz ISM band makes it very popular for WSN applications with antenna size in the range of only a few centimeters.

Center frequency	Bandwidth	Distance		
		1 m	10 m	300 m
315 MHz	0.79 MHz	22.4 dB	42.4 dB	72.0 dB
433 MHz	1.74 MHz	25.2 dB	45.2 dB	74.7 dB
868 MHz	2.0 MHz	31.2 dB	51.2 dB	80.8 dB
915 MHz	26 MHz	31.7 dB	51.7 dB	81.2 dB
2.45 GHz	100 MHz	40.2 dB	60.2 dB	89.8 dB
5.8 GHz	150 MHz	47.7 dB	67.7 dB	97.3 dB

Table 2.3: Useable frequency bandwidth and path loss under free-space conditions between ideal isotropic antennas for ISM bands defined by ITU-R as well as for additional US and European ISM bands

Figure 2.5 depicts a few popular antenna designs for WSNs. Printed circuit board (PCB) antennas are very cost effective. They need no connectors and can be balanced and matched via micro-strip technology, so they are widely used when size is in appropriate range. The disadvantages are the fixed orientation of polarization, the potentially problematic position inside the case, and the dielectric and skin effect loss of standard FR4 printed circuit board (PCB) material that becomes relevant for frequencies above around 1 GHz [17]. A commonly used antenna type is an asymmetric monopole. It has a nominal length of $\lambda/4$, but often shortened versions are utilized to save space. Their impedance is matched to 50 Ω with the help of an integrated lengthening coil. For that reason, slightly higher loss and reduced bandwidth are the consequence. Small ceramic chip antennas are used for short-range devices such as wireless local area network (WLAN) or Bluetooth because of typically reduced power efficiency.

The variety of antennas used for WSNs is high and ranges from mono- or dipole shapes, patch, or helical shape up to complex multi-band structures. In most cases, an almost isotropic radiation pattern is wanted to achieve homogeneous coverage, where the antenna gain is around 0 dBi or below, because of additional loss. In practical environment, there is no free-space condition and the radio channel is time-variant, so enough margin has to be considered to ensure

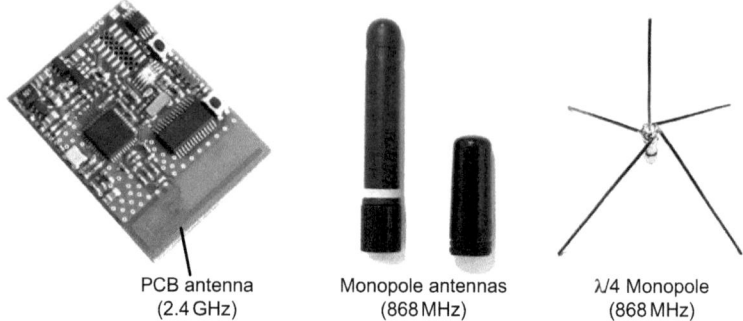

Figure 2.5: Antennas for WSN: integrated PCB antenna with balancing and matching network, a $\lambda/4$ monopole and its shorted version with integrated lengthening coil from Linx Technologies [18] and a hand-made $50\,\Omega$ monopole with high radiation efficiency, included ground plane structure for well defined characteristic, and gain of 1.0 dBi

proper operation of the whole wireless system.

2.2 Node Power Management

Strict and consequent energy and power management is mandatory for self-sufficient systems when node lifetime should exceed years [KAL05]. Application firmware has to handle all scenarios that can occur through limited amount of energy. This includes adaption of quality of service depending on actual energy reserve, and especially in case of a complete loss power supply, precautions have to be taken in order to ensure a correct power-up sequence, and restart or resume operation. In figure 2.6(a), the modified ISO-OSI reference model for communication is enhanced by a cross-layer power management plane [BFND06]. This approach allows for better optimization of overall energy consumption [MMG07]. Part (b) illustrates hardware components and features that are beneficial when minimizing system power consumption. Effective methods are sophisticated sleep modes both for CPU and peripherals with options for retention of configuration and RAM content, interrupt or event based control instead of polling, and offloading tasks from CPU via reconfigurable logic structures and periphery to reduce the number of wake-up processes for the CPU [GHG10]. A clock management unit with options for high accuracy, low power and fast start-up/wake-up oscillators offers flexibility according to actual demand. Since components of voltage supplies are powered all the time, consumption of analog circuits can dominate during low power standby modes. Various configuration/selection options of linear and switching voltage regulators can provide an optimal performance and quality balance, particularly when a combination of energy harvester and energy storage device is applied.

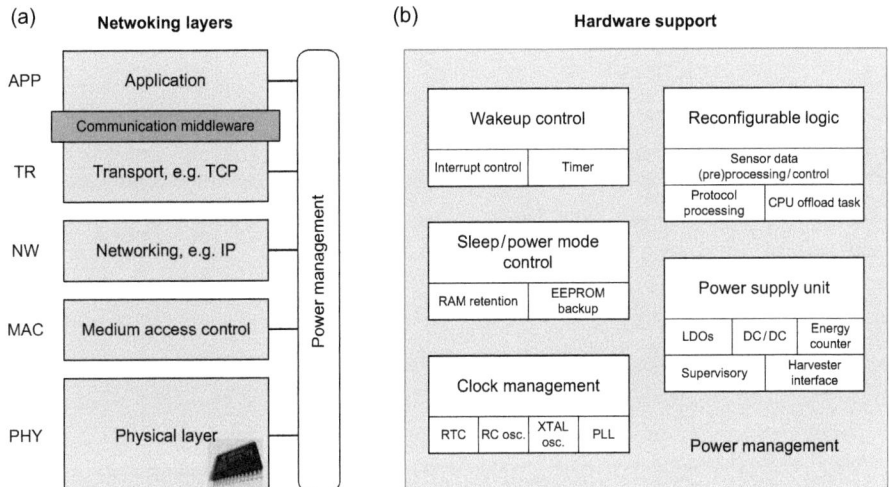

Figure 2.6: (a) Layered power management architecture of communication protocol stack and (b), hardware units supporting energy efficient application integration
LDO: low drop output voltage regulator, RTC: realtime clock, RC: resistor-capacitor, XTAL: crystal, PLL: phased locked loop

Companion chip solutions for power supply and energy management are already available for mobile handheld products from various manufacturers such as LTC3577 from Linear Technology [19]. This highly integrated power management integrated circuit (IC) includes a lithium battery charger, triple step-down switching regulators, dual low drop output voltage regulator (LDO), LED backlight control, protection and supervisory features as well as an inter-integrated circuit (I^2C) interface for control. Such ICs are not suitable for typical WSN nodes because they are designed for supplying power-hungry processors and radio transmitters in the range of up to watts and have low efficiency for light loads in microampere range. Furthermore, the amount of energy stored in batteries for portable devices is comparatively high and energy has to last only for days or a few weeks until the device is recharged. In contrast, lifetime of WSN systems is typically much longer.

An optimal power management solution would identify and get rid of bottlenecks that shorten lifetime of the sensor network in an unnecessary way. While the biggest potential for energy saving is in software and protocol design, the hardware part can be taken over by an intelligent energy management unit (EMU) with numerous options for configuration that fit to the sensor node requirements. This approach allows to balance power supply quality flexibly and dynamically according to the particular needs and the conditions during specific power states. But such EMUs for WSNs still do not exist on the market, so the problem of voltage matching between different sensor node components remains.

2.2.1 Energy Sources

Wireless sensor nodes rely on powerful energy storage or energy harvesting devices to achieve long lifetime and fulfil concurrently the given restrictions regarding size, weight, temperature range and cost. This section presents an overview of various types of energy sources and their characteristics concerning usability for WSN.

Energy Storage Devices

Batteries are very popular for energy storage because of their high energy density. They can be divided into primary and rechargeable secondary batteries. Most of all high performance batteries are based on lithium chemistry, because of the excellent energy density relating to weight [20]. Other energy storages and energy buffers are ultra capacitors, foil batteries and multi-layer chip capacitors. Characteristics and performance of state-of-the-art products are outlined in table 2.4. Nano-structured solid-state thin-film batteries have promising performance in terms of energy density as well as number of recharge cycles. The devices available today are comparatively expensive and have low capacity, but in future, high production volume, good temperature stability and the very flat structure and integrability into printed circuit boards may reduce cost enormously. Cymbet Corp. [21] and Infinite Power Solutions [22] offer SiP solutions with combinations of thin-film battery and charge management ASIC or solar energy harvesters. Table 2.4 illustrates very good storage capacity and energy density of traditional primary batteries. Nevertheless, reliability requirements under harsh environmental conditions such as high acceleration, vibration, mechanical shocks or temperature gradients necessitate alternatives. Ultra capacitors are very popular for energy buffering application, but their major disadvantages are the low nominal voltage and mainly the very high leakage current when related to storage capacity. Hence, long-term energy storage is not possible. Stored energy of multi-layer chip capacitors (MLCCs) $E = CV^2/2$, and the package volume is first of all proportional to the product of capacitance C times operating voltage V. So energy storage is more efficient at high voltage V because of the square-law characteristic, and thus results in low energy density of low-voltage capacitors. Due to the low series impedance of capacitors, combinations with batteries are widely used. This way, high battery impedance is compensated even at low temperature, where reactivity of chemistry is slow. Low self-discharge rate down to 1 % per year of modern lithium cells such as LS14250W from Saft [23] can guarantee long life-time of WSN nodes. Further important design parameters of batteries are charge efficiency, number of recharge cycles, toxicity of chemistry and charge/discharge characteristics. The assortment of battery products is high, but also requirements are versatile and heavily depend on specific application.

Type	Primary battery	Ultra capacitor	Thin-film battery	MLCC
Device	LSH20-150 (Li-SOCl$_2$) (Saft) [23]	PC10 (Maxwell) [24]	MEC101 (IPS) [22]	GRM31CR60J107 (Murata) [25]
Capacity	14 Ah	10 F	1 mAh	100 µF
Self-discharge	48 µA	40 µA	≈1 nA	55 nA
Nominal voltage	3.6 V	2.5 V	3.9 V	6.3 V
Nominal current	0.3 A	2.5 A	40 mA	≈3 A
Energy density	478 Wh/kg	1.09 Wh/kg	8.6 Wh/kg	0.07 Wh/kg
Recharge cycles	–	500,000	10,000	infinite
Temperature range	-40 – 150 °C	-40 – 70 °C	-40 – 85 °C	-55 – 85 °C

Table 2.4: Comparison of various types of energy storage devices, values given for operation at 25 °C

Energy Harvesting Devices

For many applications with lifetimes much longer than 10 years, primary batteries are not suitable because of their leakage and self-discharge rate in the range of 1 – 5 % per year and even more at elevated temperature [26]. Instead, a combination of energy harvester and rechargeable battery or energy buffer can be used. The most common types of harvesters for WSNs are:

- Solar cells: Power densities of up to $15\,\text{mW/cm}^2$ are achievable for modern solar cells at outdoor condition [27]. This is high when compared to power densities of other energy harvesters. An efficiency of more than 20 % [HSM00] for state-of-the-art cell types is acceptable. Together with simple battery charge mechanisms, solar cells are cost-effective and well established in many supply solutions.

- Vibrational scavengers: One transducer principle for harvesting vibrational energy is magnetic induction of a moving magnet or coil. Figure 2.7 depicts an application example used for powering a rail cargo monitoring system [7]. It benefits from motion of the railway axis with a frequency around 3 Hz and was developed by Linz Center of Mechatronics [28]. Because of the low resonance frequency, a big mass of 5.5 kg is necessary for an average output power of only 8 mW that is generated during one typical train run [MSD09].

IMEC [29] has demonstrated a prototype for piezoelectric energy harvesting via MEMS. It achieved an output power density of 85 µW/cm² [EPH+09].
Another MEMS harvester that exploits electrostatic principle is shown in figure 2.8. It was designed by Vestfold University College and manufactured by Infineon Norway to supply a TPMS node. It delivers an average output power of 2 µW at a driving speed of 50 km/h.

- Thermoelectric harvester: Temperature gradients together with the Seebeck effect are used to generate low voltage and harvest energy. The MPG-D715 thin-film thermogenerator from Micropelt [30] generates up to 10 mW/cm² for a temperature gradient of 10 °C. An application for it is structure health monitoring in aircrafts where temperature difference at starts and landings are utilized [1]. Other types of thermogenerators make use of body heat and are designed for biomedical application [LEVTM09].

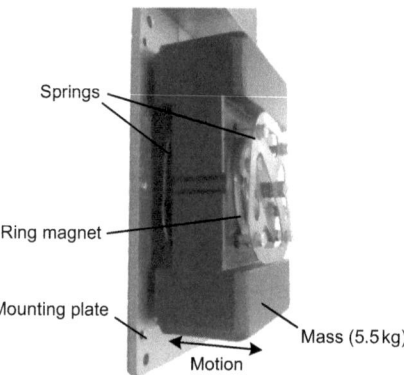

Figure 2.7: Energy harvester for wireless cargo monitoring system with magnetic transduction principle [7]

The presented energy density values for available harvesting devices look promising. To exploit almost full output energy, an appropriate power conversion interface is essential to charge an energy buffer. It has to ensure high power efficiency for the whole supply system. This includes the adequate interplay of harvester, conversion interface and power management over a large dynamic range of scavenged power. Vibrational scavengers deliver AC power with unhandy high voltage amplitudes in case of piezoelectric and electrostatic transducers. In opposite, most electro-mechanic transducers and thermogenerators produce very low voltage. Designing optimum converter interfaces that handle startup conditions also at empty energy buffers concurrently is a separate and challenging research topic where much progress is ongoing

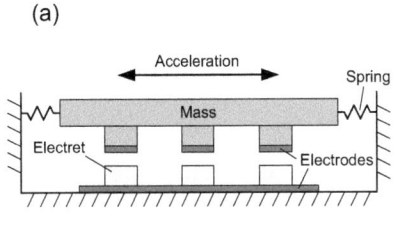 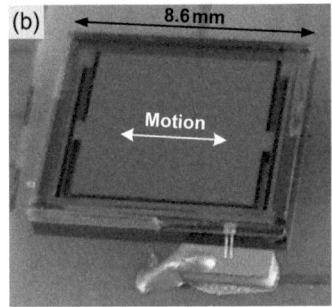

Figure 2.8: Energy harvester for TPMS from [HHJ+08]: The MEMS is mounted in-tire and utilizes acceleration peaks when touching or lifting the road. Part (a) illustrates the piezoelectric transduction principle. The harvester is precharged by an electret to 100 V and the periodic structure of the contact electrodes allow for enhanced power efficiency at little deflection due to the increased variance in electrode capacitance. Part (b) shows a die photo of a MEMS prototype that delivers up to a few microwatts of power.

in parallel to harvester development. Already available products from Linear Technology [19] are LTC3109, LTC3105, and LTC3588-1 for thermoelectric, solar, and piezoelectric transducers.

2.2.2 Power and Energy Demands

By means of a tire pressure monitoring system (TPMS) which is a typical example for a WSN application, the power consumption profile of a sensor node is analyzed. Figure 2.9 from [HHJ+08] illustrates the huge dynamic range in current consumption that is dependent from the actual state of the node. The demand in standby is below $0.2\,\mu A$ and comes from voltage regulation, voltage monitoring and leakage of electrostatic discharge (ESD) protection structures. Current consumption during pressure measurement is 1 mA for approximately 2 ms and reaches up 10 mA during data transmission phase. If a pressure measurement interval of 30 s is assumed, the accumulated charge demand for each state is $3\,\mu As$ for transmission, $2\,\mu As$ for sensor measurement, and $6\,\mu As$ during standby. Despite the very low leakage current in nano-ampere scale, energy consumption in standby state dominates overall consumption. This gets even worse at elevated ambient temperature because of the exponential nature of leakage current.

The TPMS node utilizes a multi-layer chip capacitor of $100\,\mu F$ for energy buffering. This is sufficient to store enough energy for at least one cycle of pressure measurement and datagram transmission to the vehicle's board computer (see also figure 2.2). The necessary mean power

Related Work

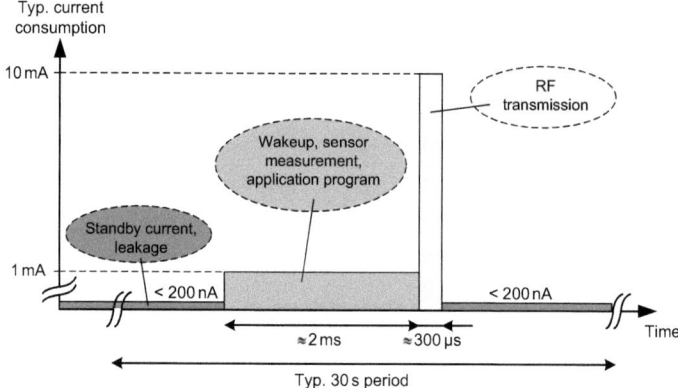

Figure 2.9: Typical current profile of a tire pressure monitoring system (TPMS) from [HHJ+08]

of approximately 1 µW for node supply is harvested via a MEMS scavenger and charges the MLCC up to 5 V. Once there is enough energy stored, the node is turned on, the microcontroller boots, does measurements and data transmission and powers down itself until a timeout wakes up the controller again, supposed that there is sufficient energy available in the storage capacitor. Then the next pressure measurement cycle starts. This radical power management concept with a full reboot process saves a maximum of energy for the tire-pressure measurement application. Nevertheless, a number of circuits are required that stay powered all the time. These are oscillator, timer, voltage monitor, voltage reference, voltage regulator, over-voltage protection circuit, an under-voltage lock-out circuit for safe and defined startup sequence, and the harvester interface for power conversion. As a consequence, the own consumption of the power management ASIC has to be clearly below the harvester's output power of 1 µW. Otherwise the system is not able to accumulate enough energy to start a new pressure measurement cycle.

From the TPMS example, one can see that always-on or mostly-on units for node supply can dominate energy consumption of a wireless sensor node. The following list gives an overview of state-of-the-art components used for power supply and power management that are relevant for design of the WuR supply unit:

- Switching voltage regulators: If high efficiency is requested for DC/DC conversion, usually switching regulators based on inductive or capacitive principle are applied. This approach is effective for load currents in milliampere scale during active node states, but inefficient at very light-load conditions or in standby operation, because most switching regulators are optimized for a high peak efficiency. DC/DC converter products with exceptionally low quiescent current are TPS60312 – a charge pump converter from Texas Instruments [9] with 1.5 µA – and XC9226 from Torex [31], which has only 15 µA consumption due to pulse-frequency modulation technique. The switching regulator's efficiency characteristic drops down for loads in microampere range mainly because of switching loss and consumption of oscillators. Thus, DC/DC converters are no suitable candidates for supplying microwatt components such as WuRs.

- Linear voltage regulators: High voltage drop leads to poor efficiency of linear regulators. This disadvantage recedes in case of light-load condition when own-consumption of LDO becomes dominant and reduces efficiency even more. In comparison to switching regulators, low own-consumption down to 0.8 µA for XC6215 from Torex [31] is available in today's products. A further advantage is the very clean supply voltage without voltage ripple, so many analog and RF circuits are supplied by LDO. An ideal solution would be a combination of a linear regulator for light loads and a DC/DC converter for high load current with automatic handover to benefit from the best of each performance.

- Voltage references: Since voltage references are needed for supply supervisory purpose, voltage regulators and partly reference current generation, they stay powered also in power save modes. Usually well-known low-power bandgap references with high accuracy and supply currents down to around 1 µA are employed [Lee09, Ann98, SBB08], but this is not acceptable when pushing requirements below some 100 nA for TPMS or WuR application. Because of limitations in chip area for poly-resistors and bandwidth restrictions, alternative design strategies for voltage references – with of course less accuracy and increased device variance – are mandatory.

- Oscillators and timers: Periodic generation of wake-up events and time measurement necessitate ultra-low power oscillators and timers. While modern quartz oscillators for real-time clocks (RTCs) with high accuracy consume down to 300 nA [32], the low frequency resistor-capacitor (RC)-oscillator from [LLC10] draws 200 nA and the 2 kHz RC-oscillator from the TPMS power management ASIC [HHJ+08] consumes only 20 nA. RC-oscillators provide fast startup, but their low accuracy can be compensated only partly via calibration techniques.

Especially at elevated temperature, wasted energy due to leakage current during long standby periods is not negligible. Beside consumption of functional supply components, unwanted leakage results from ESD protection circuits and large power gating transistors. Also self-discharge of battery is strongly temperature dependent and has to be considered when optimizing power

Related Work

consumption. If battery leakage dominates energy demand of electronics already, further effort in power reduction will not lead to significant longer lifetime.

Table 2.5 from [Mah04] shows an analysis of energy distribution over main wireless node components for typical application scenarios with a straight forward approach. Most of the total energy is consumed by radio communication. So this is an important circumstance that has to be considered both in protocol design and hardware support. An additional WuR with appropriate communication protocol can ease this bottleneck. In opposite, energy consumption of the microcontroller plays just a secondary role and is negligible if the sensor does not require extensive data processing.

Energy sink	Energy demand
Radio communication	75 – 80 %
Power supply	8 – 9 %
Microcontroller	6 – 8 %
Sensors	6 – 7 %

Table 2.5: Distribution of total energy consumption for a typical sensor node from [Mah04]

2.2.3 Medium Access Control Protocols

Communication protocol design is one of the most important issues for WSN. As seen in previous section 2.2.2, it has the biggest potential for power saving and is most responsible to operate a wireless network effectively for long time. The number of different protocols for WSN is as versatile as applications are. This section focuses mainly on aspects for power efficiency, impact on communication hardware and communication quality. In contrast to mains supplied wireless nodes, where quality of application service is the main objective, the fact of limited energy in wireless networks always leads to a compromise between quality of service (QoS) for communication and network lifetime. The most important parameters therefor are:

- Throughput: High data rate implies short utilization of the common used media and that results in low probability for packet collision. Moreover, short transmission duration saves energy per transmitted packet.

- Sensitivity: High receive sensitivity allows for long-range coverage also with moderate transmission power. Robustness against interference on adjacent channels and in-band blocking capability reduce packet error ratio and are further important aspects.

- Latency: The responsiveness of radio links is fundamental for realtime applications. Delays are accumulated even for multi-hop communication and can rapidly reach unsuitable values.

- Complexity: Sophisticated protocols require knowledge about the network such as timing information, topology, energy reserve, or routing address information. Mobile nodes can join or leave a network cluster and all information has to be gathered and exchanged during each setup or reconfiguration phase. Low complexity helps to establish communication links fast and saves power due to sparse overhead. Even when excessive duty-cycling of transceivers is applied, synchronization of time-slots may take long time and much energy.

- Power consumption: As presented in table 2.1, data transmission and reception have approximately the same cost in terms of power consumption. So strategies for power saving are mandatory for both modes of operation. Much effort is taken in protocol design specifically for MAC and routing layer to push the mean power demand of around 25 mW for transceivers well down to microwatt range. The consequence is a large amount different of protocols. Each is designed for a certain class of applications with varying tradeoff between performance and power consumption.

MAC protocols can be divided into two categories, scheduling and non-scheduling protocols. Scheduling strategies make use of a technique for time scheduled synchronization of network nodes. The whole network or at least a cluster of nodes communicate during well defined periodic time slots. So the receiver can be shut down for the rest of the time interval and it has to listen for incoming traffic during these short time slots only. This time division multiple access (TDMA) approach saves lots of energy for small duty-cycles. If for instance a receiver consumption of 25 mW and an on-time to period ratio of 1/10,000 are assumed, the mean power consumption would reduce to 2.5 µW with the penalty of increased latency. Also for the optimal case with high data rate and short packets, when the actual period for data transmission can be neglected, a minimum remaining on-time is required for startup of the oscillator circuit. This is typically 1 ms for conventional receivers (see table 2.1) and leads to a scheduling period of 10 s, if the small duty-cycle of 1/10,000 is considered and thus, network latency of up to 10 s is the consequence already for a single hop. Network reactivity drops down rapidly to unsuitable values even for multi-hop links. A further disadvantage of excessive TDMA is the increased probability for packet collision during the short time slots, where all network traffic is concentrated. Adaptive scheduling intervals can help to improve network reactivity and reduce the collision probability on demand, but complicate protocol design. Beside latency, node synchronization is a big and power hungry issue because probability to hit a short dedicated time slot randomly within a long period is very low, or the receiver has to listen for at least one complete TDMA period of 10 s to guarantee a hit. A principle limitation in duty-cycle is given through the clock drift of independent node clocks. Standard low-cost realtime clock (RTC) oscillator crystals such as X3215 series from Euroquartz [33] have frequency tolerance of more than 100 ppm over the full temperature range of -40 – 85 °C. Absolute tolerance and aging effects can be compensated via calibration techniques. Frequency deviation caused by temperature drift of commonly used 32,768 kHz quartz crystals has parabolic temperature characteristic. Its zero point is shifted nominally to 25 °C for X-cut vibration mode. The quartz material specific temperature characteristic of -0.04 ppm/°C^2 for X3215 is well known a priori. So it is possible to further extend accuracy and duty-cycle down to around 1 ppm with the help of temperature compensated oscillators such as EM57T from Euroquartz [33]. Then clock accuracy is no more a bottleneck for design of typical MAC protocols. Nevertheless, synchronization and re-synchronization mechanisms have to be applied to establish and hold radio link connections. Dynamic clock adjustments ensure re-synchronization and compensation of drift and jitter and lead to necessarily broadened time slots with some overlap to guarantee that all receivers listen to the radio channel already before the actual data is transmitted.

In opposite to scheduling protocols, non-scheduling MAC protocols do not use time information for synchronization of communication. So the receiver has to listen for arriving data packets all the time for the most simple approach. Therefore, the receiver from calculation example would consume 25 mW which is not suitable for low power WSN applications. Practical implementations make use of combinations of both scheduling and non-scheduling strategies with various extensions and optimizations.

The CSMA-PS protocol [EH02] was proposed in 2002 and uses a preamble sampling technique. The duty-cycled receiver periodically listens to the radio channel for a preamble from an unsynchronized transmitter. To guarantee a hit by the receiver, a preamble length of at least one period of the receiver's duty-cycle is necessary before actual data transmission can start. Once the preamble is detected, the receiver stays active until data content is transferred. This approach allows flexible trading of power versus latency, dependent on duty-cycle. The synchronization process is done each time before communication phase and costs comparatively much energy because of the long preambles.

The STEM protocol from [STGS02] is similar to CSMA-PS, but uses an additional wake-up radio channel that allows a reduction of preamble length in average and thus saves power and latency. WiseMAC [EEHDP04] exploits knowledge of time schedule information of neighboring nodes to shorten wake-up preamble and overhead via a prediction technique. A combination of STEM and WiseMAC features leads to CSMA-MPS [MB04], a minimum preamble sampling MAC protocol with further power optimization. PMAC [KB06] is a powerful and complex protocol with good network performance. It makes use of many different strategies for perfor-

mance and power optimization, but needs a separate bootstrap phase for network setup.
The MAC layer defined in IEEE 802.15.4 standard for instance is designed with focus on energy management and simplicity instead of data throughput. In beacon mode, it makes use of super frames with well defined time slots that are managed by the coordinator node. In non-beacon mode, communication happens asynchronously via CSMA/CA method. IEEE 802.15.4 is used for the ZigBee [10] and the 6LoWPAN [34] protocol stack.
It can be concluded that the more performance optimization is done, the more knowledge about network structure and status is necessary for operation. So information gathering becomes expensive even for ad-hoc networks with joining and leaving network nodes.

The way out of the latency versus power demand dilemma is an approach with WuR based MAC protocols [DEO09]. The additional and dedicated ultra-low power receiver has to listen for wake-up preambles only. This can be done all the time without scheduling because of low power demand. The preamble is added in front of the conventional data packet header. It contains special data content that is matched to the WuR's needs as well as address information that is used to wake-up specific wireless nodes and includes individual, multi-cast or broadcast addresses. Once the WuR detects a wake-up event and an address match within the received preamble, the sleeping main receiver is woken up and a traditional communication cycle starts. The benefits of this approach are a very simple extension of standard communication protocols, flat distribution of channel utilization over time and thus low probability for collision. There is no need for any kind of time prediction or synchronization of time slots. So latency is reduced to a minimum in spite of microwatt power consumption and furthermore, a setup phase for gathering network information is obsolete.

In [TSM+09], a wake-up module with power demand of 12.4 µW is utilized to construct a basic communication scheme with low-power idle listening. The WuR waits for dedicated wake-up packets from the sender and notifies the receiving data communication module via a wake-up signal. In a second step, the data packet is then transferred from the sender's data communication module to the main receive module. The wake-up packet includes a preamble that initializes the analog circuitry first. Then it is followed by an ID code that is utilized as node address. The wake-up hardware module consists of two main stages and is depicted in figure 2.10. A simple radio wave detector discovers the presence of a wake-up preamble first, and secondly starts the demodulator and decoder that checks ID matching of the wake-up packet. This approach saves energy, because of gating of the power hungry address decoding part that is active only if a preamble of sufficient signal strength is detected first.

Comparison between preamble sampling techniques in MAC protocols and the WuR based approach is presented in [SWP10]. With the help of off-the-shelf transceivers and two WuR designs, the benefit of latency reduction due to the WuR is clearly shown for scenarios with

Related Work

comparatively low energy consumption. However, the main disadvantage of 20 – 30 dB lower receive sensitivity when compared to commercial receivers is pointed out and leads to significantly reduced coverage in practice.

In [NT02], scheduled rendezvous technique is compared with an RFID wake-up receiver based architecture. In situations with low network utilization and high level of responsiveness, the low power wake-up can achieve much lower power consumption than the scheduled approach. The authors suggest a more complex hybrid protocol by combining benefits from each concept. That would allow lowest power consumption for a given latency requirement on basis of the best compromise.

2.3 Wake-up Receiver

There exist various approaches for implementing a low power wake-up receiver in literature. Architecture, design strategies and performance parameters of most promising concepts are compared and discussed in this section.

2.3.1 Concepts and Performance

The main requirement for competitive wake-up receiver architectures and solutions is low power consumption well in the range of microwatts. WuRs need to listen to the radio channel for incoming packets all the time to ensure high node reactivity and low latency. As a consequence, they are normally not duty-cycled and thus, always-on operation with low current consumption is mandatory.

A simple approach for a wake-up module is suggested by von der Mark et al. in [vdMKHB05] and Takiguchi et al. in [TSM+09]. It is illustrated in figure 2.10. The process for wake-up event detection is divided into two steps with the help of two stages. The first one consists of a radio wave detector that simply scans for a minimum level of RF amplitude of an ASK/OOK modulated RF carrier. Once a certain threshold is exceeded, the power controller turns on the second stage that performs address decoding afterwards. It consumes significantly more power than the first stage mainly because of the LNA for sensitivity gain and the more complex ID match detection unit that finally generates the wake-up signal. This power gating principle saves much energy in the normal case of low probability for radio traffic. Since limitation of receive sensitivity is given by the RF detector of the first stage due to lack of an amplifier, the authors propose to transmit a short high-power wake-up burst to activate the power controller first and reduce transmit power subsequently during ID transmission phase for the second stage of the wake-up module. This principle seems to be power efficient, but in practice, the transmit power is limited by the transmitter's supply voltage of typically not more than 3.3 V. Without a special power amplifier, common RF output stages have an open drain transistor that is connected to the matching network and biased with supply voltage. Further boost of RF power would require an additional power amplifier together with a voltage converter. The consequence of low receive sensitivity is reduced radio link coverage when compared to state-of-the-art main receivers. A very similar architecture based on the two stage wake-up mechanism with receive signal strength detection and subsequent address decoder is used in [vdMB07].

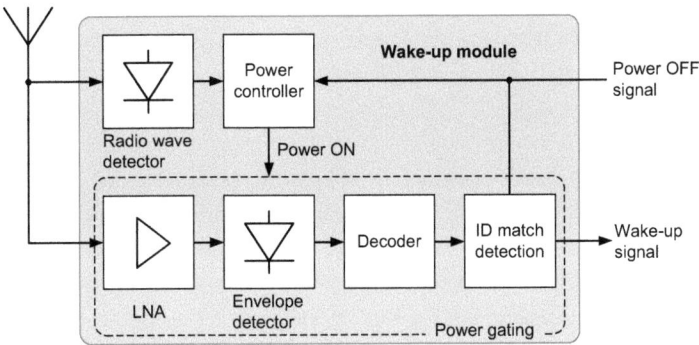

Figure 2.10: Two stage wake-up receiver concept from [TSM+09], [vdMKHB05] and [vdMB07]
LNA: low noise amplifier

One main difference between the mentioned designs is the implementation of the RF detectors. In [TSM+09], the radio wave detector also illustrated in figure 2.11(a) is fully CMOS integrated and utilizes two n-channel metal-oxide semiconductor (NMOS) transistors. Their gate-source voltage is zero, so the bulk diodes are used to construct a peak voltage rectifier. C_2 filters out RF carrier and the demodulated baseband signal is provided at LF_{out}. This approach is plausible and the power consumption of 12.4 µW seems to be promising, but there is no measured or simulated result for receive sensitivity given. The estimated value of only -36.9 dBm is used for exemplary calculation and allows a link distance of just a few meters at optimal conditions even without barriers.

The envelope detector in figure 2.11(b) from [vdMB07] consists of Schottky diodes and is integrated in 250 nm SiGe BiCMOS technology for 2.4 GHz operation. Together with the following discriminator that is constructed of three inverter stages, it represents the main part of the first WuR stage from figure 2.10. The diodes with high saturation current form a Villard cascade to multiply the input voltage level and thus gain sensitivity. The main benefit of the design is low power consumption due to the absence of bias current for the detector diodes. Nevertheless, without bias current for the Schottky rectifier, the input impedance of the RF frontend is extremely high. So it is hardly possible to match the antenna's commonly 50 Ω impedance with low loss in the matching network and with low reflection coefficient. A further weak point of the presented detector design is frequency selectivity. With exception of the mentioned antenna matching network, there is no frequency determining RF component in the whole design. The broadband envelope detector can cover a large frequency range between some hundred megahertz up to a few gigahertz. This offers flexible application on one hand, but on the other hand, wide bandwidth without frequency band selection collects lots of interference. Especially in 2.4 GHz ISM band, problems with WLAN, Bluetooth and ZigBee devices can cause unwanted wake-up events that trigger the second WuR stage and lead to increased power consumption. In [vdMB07], the demodulated bit sequence is used for address match detection. Already a single bit error caused by interference leads to an address miss match and thus a missed wake-up event. So high signal-to-noise ratio (SNR) is required to ensure low bit error ratio (BER). The performance of the proposed discriminator stage in figure 2.11(b) with the main task of signal amplification and quantization was only simulated. Parasitic effects at analog usage of CMOS inverters such as flicker noise, thermal noise, thermal drift and variability of transistor threshold voltage as well as subthreshold current are not mentioned nor considered, and would degrade performance dramatically. So it is hard to believe that the given detector sensitivity of -50 dBm would be achievable in practice without demonstration of an ASIC implementation. Furthermore, the impact of interference would degrade performance significantly in real world environment.

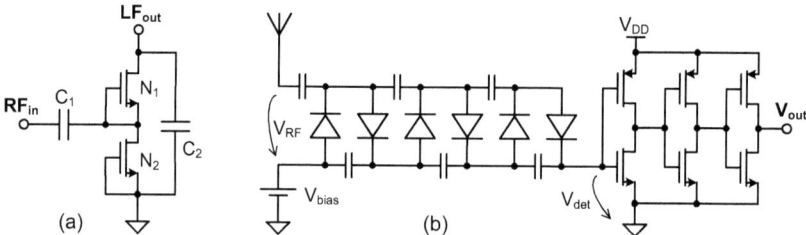

Figure 2.11: Radio frequency detectors: (a) MOS detector from [TSM+09] and (b), Schottky detector with zero bias current and inverter based discriminator from [vdMB07]

A diode detector for wattless reception is reported in [MAB03]. The paper focuses on optimized design of a patch antenna for high output voltage and high impedance load to increase sensitivity of a single rectifier diode. The discriminator is more or less the same than those from [vdMB07] discussed before in figure 2.11(b) with all its problems and difficulties in robustness for practical application. Beside negligible power consumption of the detector itself, demand of signal conditioning circuitry is not considered in this concept for wattless wake-up.

The wake-up receiver architecture proposed by Le-Huy et al. [LHR08] is illustrated in figure 2.12(a). It contains an envelope detector frontend for ASK/OOK modulated carrier signals at 2.4 GHz and an address decoder unit. The core of the decoder is a pulse width modulation (PWM) demodulator stage that extracts the serial bit stream before it is compared to the node's address code for match detection. This architecture operates without a power gated low noise preamplifier in front of the RF detector and thus, adequate design for high sensitivity is very important. The envelope detector drafted in figure 2.12(b) relies on a voltage doubler circuit with zero-biased Schottky diodes. Devices C_1 and L_1 build the matching network for $50\,\Omega$ antenna impedance and C_2 filters out the residual radio frequency. Again, low-loss power-matching is hard to reach because of the diode's high differential resistance. A WuR power consumption of $20\,\mu W$ is estimated for the whole design. This covers only baseband signal evaluation because of the zero-power RF detection approach. The receive sensitivity of around -50 dBm is just stated without detailed information about conditions such as SNR, BER, wake-up detection probability or false wake-up rate (FWR). The simulation result of detector sensitivity versus RF input power equates theory for medium and high input levels, but the deviation from square-law at low amplitudes is not explained and degrades relevance of the results, since there is no physical realization.

The WuR architecture from Marco Spinola-Durante was developed at the Institute of Computer Technology at Vienna University of Technology during the course of his PhD thesis [SD09] and is very similar to figure 2.12(a). The design contains of an envelope detector based radio frontend with a single off-chip Schottky diode and a PWM demodulator stage that converts the desired baseband signal to digital domain. Afterwards, an enhanced digital signal processing unit with forward error correction capability minimizes the residual bit errors which are a consequence of noise. Thanks to error correcting codes, the WuR is able to accept reduced SNR. So the receive sensitivity is increased up to -57 dBm [SDM09], but this is still very low for practical application also for short radio links. An estimated overall power consumption of $12.5\,\mu W$ is on the other hand quite good.

An interesting approach with high receive sensitivity is reported in [KL07] and depicted in the block diagram of figure 2.13(a). A passive analog RF frontend with narrow IF bandwidth offers high signal-to-noise ratio and high dynamic range. Without the need for active signal processing in RF domain, this concept has inherently high potential for low power consumption. The subsequent IF amplifier and baseband detector operate at 455 kHz and thus can be implemented with low power demand. Finally a message decoder is utilized for detection and signaling of a wake-up condition similarly to previous discussed WuR designs. Figure 2.13(b)

Related Work

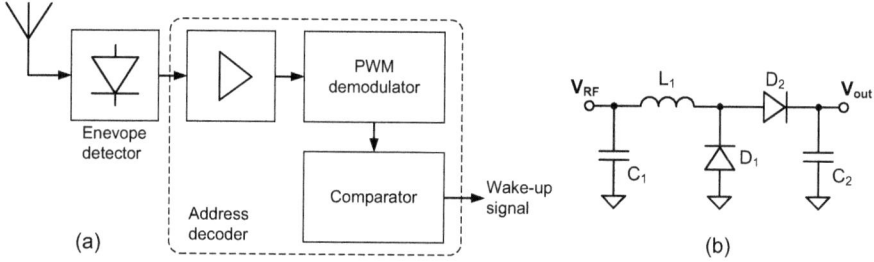

Figure 2.12: (a) Wake-up receiver architecture and (b), envelope detector based on zero-bias Schottky diode voltage doubler circuit with matching network from [LHR08]

shows the passive RF frontend. It includes a 916 MHz SAW filter for frequency band selection and out-of-band interference suppression as well as a matching network for the zero-biased Schottky detector with differential resistance of 8.2 kΩ. The innovative principle is signal conversion to an intermediate frequency of 455 kHz instead of direct down conversion to baseband domain. The IF filter has a 3 dB bandwidth of only 6 kHz, filters out most of the noise, and suppresses interference. Therefore it gains SNR, receive sensitivity and improves dynamic range. To benefit from this design, a more complex modulation technique is necessary what limits applicability. The 916 MHz carrier frequency is first of all OOK modulated with periodic "01" bit sequence at 910 kbit/s. The resulting signal is then modulated again with a bit stream that contains the actual transmit data. One disadvantage of this modulation scheme is the high data rate of the transmitter that delimits assortment. Another drawback is the occupation of radio channel bandwidth of at least 1 MHz and this is not available even in the European 868 MHz ISM frequency band. The reported tangential sensitivity of -69 dBm at 8 dB SNR is excellent for the anticipated power consumption of 20 µW for IF signal processing. However, the SAW filter, the IF filter and the transformer used for IF impedance matching are very bulky external devices and prevent high integration density and low system cost.

In [PGR07] Pletcher et al. present an RF to digital baseband WuR for 1.9 GHz operation that is realized in 90 nm CMOS. The analog frontend is depicted in the block diagram of figure 2.14(a). It includes a frontend amplifier (FEA) with matching network based on a BAW resonator for low noise RF gain, and is followed by an envelope detector that is realized via MOS transistors operating in subthreshold region. This approach of detector implementation allows for efficient nonlinear conversion also with MOS transistors. Finally the analog baseband signal is amplified by a programmable gain amplifier (PGA) and converted to digital domain with a resolution of 6 bit. Digital signal processing is implemented off-chip with the help of the MATLAB software tool on a personal computer. Figure 2.14(b) shows the input stage of the frontend amplifier (FEA). Application of a BAW resonator together with the integrated RF

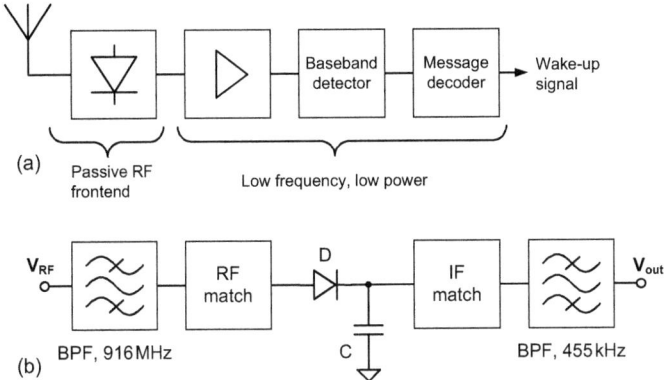

Figure 2.13: (a) Wake-up receiver architecture and (b), passive RF frontend from [KL07]
BPF: bandpass filter

matching network consisting of C_1 and C_2 eliminate bulky SAW filters and external inductors and provide small bandwidth for high sensitivity and high robustness against interference. The FEA stage provides 10 dB of gain and 7 MHz bandwidth via the cascode stage N_1, N_2. The high amplifier load impedance is generated via an active inductor simulation network via P_1 and P_2.

The major power consumption of 48 µW is spent to the FEA to achieve high sensitivity. Envelope detector, PGA, analog-to-digital converter (ADC), and bandgap reference (BGR) together consume only 17 µW. When detecting a 31 bit code sequence, the WuR exhibits -56 dBm sensitivity at 100 kbit/s, 90 % detection probability, and a noise-caused fail alarm rate of 1/s. Overall power demand is 65 µW from a 0.5 V supply when consumption of off-chip signal processing is excluded. The low supply voltage helps for reduction of power consumption, but is unhandy for integration into embedded systems. Efficient step-down voltage conversion with DC/DC regulators requires low relative voltage drop at the switching elements as well as low own-consumption of clock generating oscillators that even becomes significant at the light load of WuRs. The alternative of a linear voltage regulator suffers from high voltage drop in case of a typical battery voltage of 3 V, so power supply circuitry can have crucial impact to system efficiency. This WuR implementation benefits from sensitivity enhancement due to off-chip digital signal processing, but still has too low sensitivity to cover radio link distances of more than a few meters with suitable reserve in the link budget. The measured noise figure of 9.5 dB indicates further margin of sensitivity improvement.

A different approach for the RF frontend is presented in [PRG09, Ple08] and outlined in figure 2.15. In opposite to receiver architectures discussed before, the authors propose a super-

Related Work

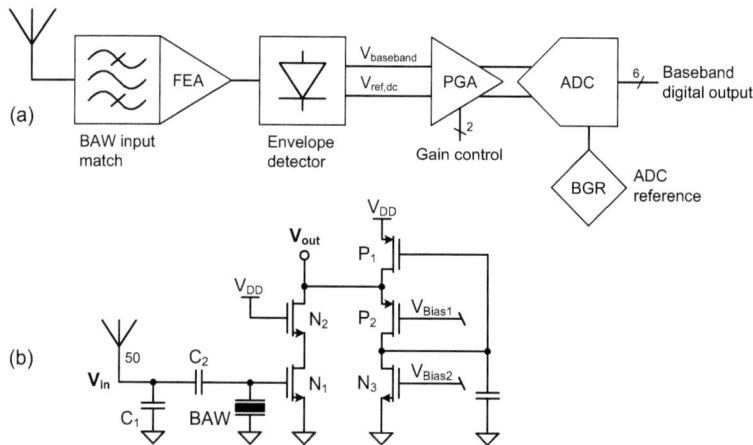

Figure 2.14: (a) Wake-up receiver architecture and (b), RF frontend amplifier (FEA) with BAW resonator input network from [PGR07]
BGR: bandgap reference, BAW: bulk acoustic wave

heterodyne principle with uncertain IF architecture. The WuR is fabricated in 90 nm CMOS technology and designed for OOK modulated signals in a 2 GHz frequency band. Its block diagram is shown in figure 2.15(a) and includes a BAW based pre-selection filter that is integrated into the input matching network with approximately 12 MHz bandwidth. A dual-gate mixer converts the filtered RF input signal down to intermediate frequency (IF) domain that ranges from 1 MHz to 100 MHz, before wideband IF amplification. The envelope detector removes phase and frequency content of the IF signal and converts the signal's envelope to output amplitude in baseband domain. Most power of the entire WuR system is consumed by the local oscillator (LO) that generates frequency for mixing. To save power, this key component is implemented as free-running ring oscillator and embedded in a frequency regulation loop. The digitally controlled oscillator (DCO) consists of three CMOS inverters with digitally adjustable supply voltage for LO frequency tuning. Calibration of frequency is done off-chip and is necessary because of temperature dependent drifts and long term effects. Figure 2.15(b) illustrates the frequency plan of the WuR architecture. Thanks to the BAW based high-Q frontend filter, RF channel selectivity is quite good. In worst case, the LO's frequency may drift by ±100 MHz around the desired 2 GHz carrier frequency in spite of calibration. The consequence is an uncertain IF architecture from 1 MHz to 100 MHz to ensure that the down-converted receive signal is covered by IF range in any case. Finally the envelope detector converts the full IF band to baseband domain. The implemented receiver frontend occupies only 0.1 mm^2 of semiconductor area without the external BAW resonator. It achieves high sensitivity of -72 dBm at 100 kbit/s data rate and a BER of 10^{-3}. The power consumption is 52 µW from a 0.5 V supply.

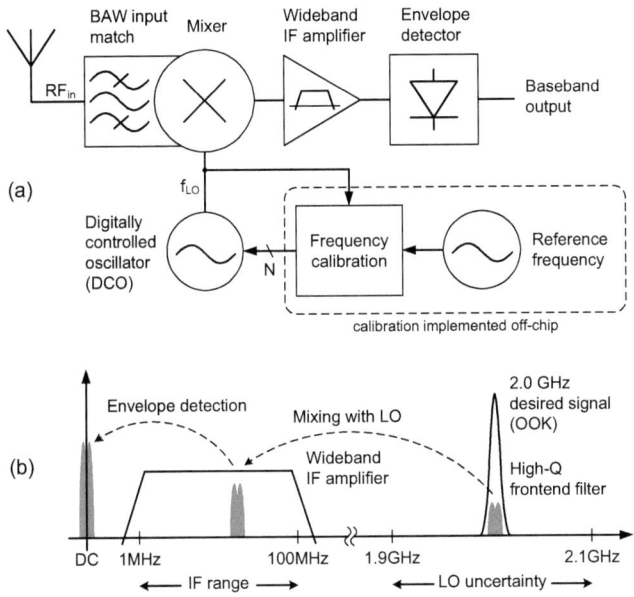

Figure 2.15: (a) Wake-up receiver block diagram of analog frontend and (b), frequency plan of uncertain-IF receiver operation method from [PRG09]
IF: intermediate frequency, LO: local oscillator

In contrast to the RF envelope detector based frontends presented before, this approach benefits from inherent linearity of RF amplitude conversion via the superheterodyne principle. Therefore, the architecture has high potential for good sensitivity and increased dynamic range. For example, 10 dB reduced RF signal strengths leads to 10 dB lower output amplitude and thus 10 dB inferior baseband SNR for linear systems. In case of analog frontends based on square-law detectors such as [PGR07], a 10 dB step results in 20 dB worsened SNR at the baseband output. So the simple detector approach lacks in receive sensitivity because of a rapid drop-off in SNR with decreasing signal strength. On the other hand, power consumption of local oscillator and mixer is already 28 µW and prevents a WuR implementation below the targeted power level of 5 µW. Principally, the frequency conversion architecture would allow flexible radio channel selection, but the mandatory BAW resonator used for channel selection, blocking of adjacent interferers, and suppression of noise and image frequency has the clear disadvantage of a fixed frequency. The large IF bandwidth of 100 MHz cannot be used for further reduction of the down-converted noise, but it is necessary because of the frequency drift and phase noise caused by the ring oscillator. A LO frequency with higher accuracy and lower drift can lead to further enhancement of receive sensitivity when IF bandwidth is reduced concurrently to a few megahertz. The penalty would be a more complex frequency calibration technique for the LO frequency generating DCO. The architecture exploits image frequency of mixing cleverly to map LO frequency tolerance of ±100 MHz to 0 – 100 MHz in IF domain. It is unlikely that the LO frequency is equal to the desired RF carrier frequency over long time because of immanent phase noise and drift. But if so, the receiver would not work because the resulting difference frequency is zero and thus would be blocked by the IF passband filter.

Figure 2.16 presents a receiver frontend from [HRW+10]. It utilizes a double-sampling technique to reduce performance degrading influence of flicker noise that is usually present in analog CMOS circuits that operate at low frequency. The 90 nm WuR ASIC is developed by Huang et al. at IMEC, Belgium. The block diagram shows a common envelope detector based analog frontend with impedance-matched low noise preamplifier and baseband amplifiers. The innovative parts are the additional input switch and the output sampler stage. By narrowing bandwidth via reduction of data rate, noise power decreases and thus receive sensitivity increases for a given bit error ratio. This relationship is commonly utilized by low power receivers. At frequencies below the $1/f$ corner, spectral noise density of flicker noise dominates over the thermal noise of CMOS devices. Data rate reduction then becomes less effective to gain receive sensitivity. Consequently, this implementation focuses on suppression of $1/f$ noise and offset cancelation. The switch in front of the LNA alternates between antenna input and ground, while the baseband output is sampled each time to distinguish between pure noise caused output signal and desired signal. Frequency domain behavior is illustrated in figure 2.16(b). The sidebands of the RF input signal at $f_C \pm f_{CLK}$ are the consequence of toggling the input switch. The envelope detector then converts RF signal to f_{CLK} and a passband filter eliminates the resulting harmonics. The sampling frequency f_{CLK} is 10 MHz and located right above the $1/f$ corner of flicker noise. The amplified and filtered baseband signal is sampled back to DC via the output sampler. Concurrently, the flicker noise is up-converted to multiples of f_{CLK} and filtered out finally. This way, the presented receiver eliminates nearly all $1/f$ noise contribution and achieves sensitivities of -75 dBm at 915 MHz or -64 dBm at 2.4 GHz for 100 kbit/s and 12 dB of SNR. The excellent performance also results from the high LNA gain of 26 dB and an external matching network with high-Q inductor for high voltage gain. Without bond pads, the net demand of chip area is 0.36 mm^2.

This WuR is externally supplied with a clock of 20 MHz. Power consumption for clock generation is not considered in the overall power budget. However, performance and especially receive sensitivity are already suitable for practical application, but a total power consumption of 51 μW from 0.5 V supply is still very high for button cell powered sensor nodes. A power

Related Work

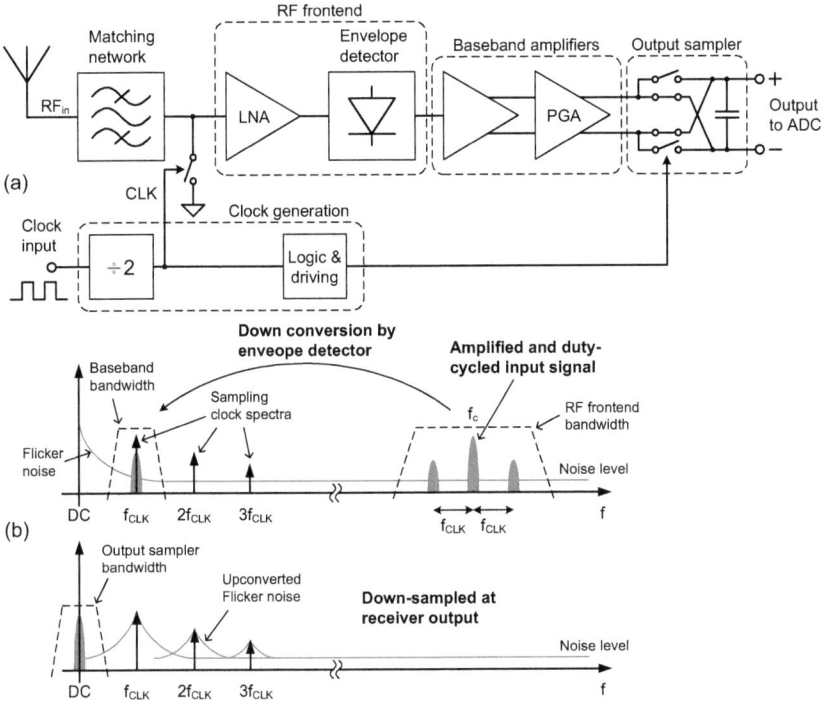

Figure 2.16: (a) Block diagram of wake-up receiver frontend based on double-sampling technique and (b), frequency domain illustration of double-sampling envelope detector from [HRW+10]
PGA: programmable gain amplifier

consumption of 27 µW for the low noise amplifier is not amazing for this receiver architecture. Beside the matching network, there is no further part for frequency band selection, so immunity against interference on adjacent channels is very limited.

A very different approach for a low power WuR frontend that is based on a super-regenerative oscillator principle is presented in [XJSCSG08] and depicted in figure 2.17. The 180 nm CMOS design for OOK modulated carrier frequencies in the 2.4 GHz ISM band makes use of startup time dependency $t_{startup}$ of an LC oscillator from an initial oscillation amplitude A. When τ is the time constant of an abating or raising oscillation of an LC resonant circuit, startup time $t_{startup} = \tau \ln(V_{OSC}/A)$, where V_{OSC} is a certain amplitude threshold. Under typically linear condition in small signal domain, a for example noise-caused initial oscillation amplitude is amplified and grows exponentially until nonlinear effects clip the output amplitude. So oscillator startup time is an extent for inference to the initial amplitude and thus receive signal strength. Periodic quenching of oscillation and measurement of $t_{startup}$ is then comparable to sampling of the RF channel amplitude. Evaluation of startup time differences allows for decoding of OOK modulated antenna signal. The additional isolation amplifier in figure 2.17(a) prevents transmission of forced oscillation signal when propagating back to the antenna, and the envelope detector determines the amplitude threshold for termination of time measurement. A quenching signal of 100 kHz equates a 100 kbit/s data rate and gates the power supply of the analog part via a duty-cycle of 10 % to reduce total power consumption to 56 µW from 1.8 V supply. Figure 2.17(b) illustrates the simulated performance results. The linear time dependency from logarithmic receive signal strength at low input power clearly shows exponential characteristic from equation above, while at high RF power levels, nonlinear saturation effects get visible. A receive sensitivity of -75 dBm is expected but hard to believe for physical chip realization, since the results are simulated only and the presence of noise is neglected completely in all considerations. Without noise, the presented results indicate even better sensitivity beyond -100 dBm, but in physical environment noise would for example start the oscillator earlier than expected, or even stop low amplitude oscillation in case of extinction. This finally leads to extensive jitter for time measurements. Furthermore, bandwidth of LC oscillator also with the proposed high quality bond-wire inductors is high and thus, sensitivity to interference on adjacent radio channels is problematic for application.

In [DLS+10], the authors describe a crystal-less wake-up receiver for the 2.4 GHz frequency band without external components for RF and IF filtering. The 65 nm CMOS ASIC is designed for pulse position modulation (PPM) and demonstrates a fully integrated receiver solution with very low space requirement of just 0.2 mm². This design is driven by an external 10 MHz reference clock with relaxed accuracy of 0.5 %. So small ceramic resonators can be used instead of bulky quartz crystals and this additionally supports low cost wireless nodes. The receiver architecture relies on a combination of superheterodyne architecture and non-coherent energy detection principle at the broadband-IF domain. Figure 2.18(a) shows a block diagram of the WuR frontend. The receive chain in the upper half consists of input matching network, image rejection mixer, wideband IF amplifier stage and full-wave rectifier, while generation of LO frequency is ensured by a digitally controlled oscillator (DCO) together with a finite state machine and the external reference clock for frequency calibration. Since local oscillator dominates power budget, it is duty-cycled together with the complete RF frontend in order to reduce power consumption. Periodic calibration of the free-running DCO eliminates temperature and supply variation caused frequency drift in a frequency control loop. The RF input filter/match includes a passive LC resonator with on-chip inductors for interference suppression. Quadrature outputs of the local oscillator, image rejection mixer with in-phase component (I)

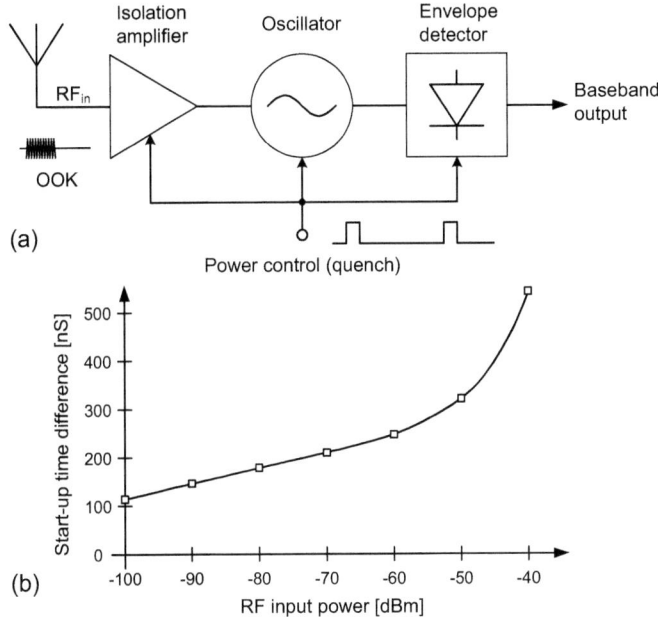

Figure 2.17: (a) Block diagram of super-regenerative wake-up receiver frontend from [XJSCSG08] and (b), simulated oscillator startup time versus receive signal strength
OOK: on-off keying

and quadrature component (Q) outputs, and the subsequent polyphase filter attenuate the unwanted image signal in IF domain. Finally, the rectifier detects energy content of the incoming data pulses and converts them to the baseband output. The time domain scheme of pulse position demodulation is illustrated in figure 2.18(b). The burst length of the RF input signal is 80 ns at 500 kbit/s. To detect position of the input signal, the DCO output is active during both possible alternatives of a binary PPM scheme. An extended DCO burst length of 100 ns allows sufficient margin for signal stabilization and guarantees proper down-conversion to the IF first, and to the baseband output later on. Synchronization of receiver and transmitter for suitable PPM decoding is ensured off-chip. The frequency plan in figure 2.18(c) shows a nominal carrier frequency of 2.44 GHz. With the help of the 2.39 GHz ±0.5 % LO frequency, the quadrature mixer accomplishes first down-conversion. Due to this frequency tolerance, the intermediate frequency after mixing lies somewhere between 26 MHz and 74 MHz and the full-wave rectifier manages the second down-conversion to baseband domain. This non-coherent energy detection makes it possible to tolerate IF uncertainty. Thanks to the image rejection mixer, the image

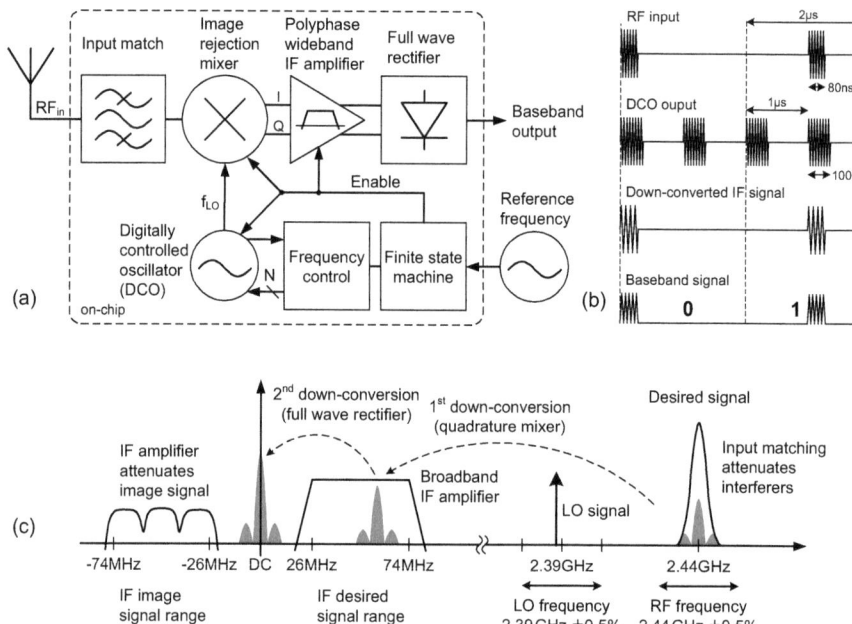

Figure 2.18: (a) Block diagram of wake-up receiver frontend from [DLS+10], (b) time domain illustration of duty-cycled down-conversion principle for pulse-position modulated RF input and (c), corresponding frequency domain visualization
I: in-phase component, Q: quadrature component, IF: intermediate frequency, LO: local oscillator

signal in IF domain is attenuated. So overall noise figure is reduced further, and robustness against interference increases. A measured sensitivity result of up to -87 dBm at 250 kbit/s and a BER of 10^{-3} is excellent.

However, this WuR implementation consumes indeed 415 µW of power because of its high data rate, but the reported energy consumption of only 830 pJ/bit is promising at -82 dBm sensitivity. Principally, this receiver architecture scales linear with data rate and thus, power demand can be reduced to below some 10 µW at a data rate of few kilobit per second, which is sufficient for generation of wake-up events. A clear disadvantage of this realization is its large occupancy of RF bandwidth. The ±0.5 % tolerance of the reference frequency and DCO, and also the large bandwidth of the PPM pulses require a radio channel bandwidth of at least 48 MHz. Consequently, narrow-band application in very limited ISM frequency bands below 1 GHz is impossible.

Figure 2.19 presents performance of state-of-the-art WuR designs as well as characteristics of a few selected low-power main receivers. The chart shows the major tradeoff between power consumption versus receive sensitivity for WuR application and gives a rough performance overview. Not all environmental conditions for the presented values are described fully in literature and some of them are missing completely, so it is not possible to compare the given results exactly also because of the very different receiver architectures or varying data rates and modulation schemes. Nevertheless, a survey of suitability is given and promising candidates can be identified. All designs have desired input frequencies in the range from 868 MHz to 2.4 GHz, but not all of them have been integrated fully into an ASIC solution. Some designs have not been fabricated and thus, the presented results rely on simulation and estimation. The main issue is to achieve high receive sensitivity and low power consumption at the same time and push current limitations towards lower barriers. This is mandatory to avoid a bottleneck in radio link budget when compared to main receivers such as TDA5240 from Infineon [4]. Its very high sensitivity of -119 dBm at 2 kbit/s and FSK modulation would be useless if it cannot be woken up by the WuR device due to lack of enough sensitivity. The design with lowest power consumption of 12.5 µW from [SDM09] was developed at the Institute of Computer Technology. It uses an external Schottky diode detector with enhanced digital signal processing in baseband domain that includes a forward error correction scheme. But the measured sensitivity of -57 dBm is very low for practical application. The design from Berkeley Wireless Research Center [PRG09] has acceptable sensitivity of -72 dBm, but power consumption is already 52 µW, and designs with excellent sensitivity better than -90 dBm consume unsuitably high 400 µW.

With the exception of only a few publications, most literature of good performing low power WuR designs is published during the last two years. This fact indicates major interest and necessity of the WuR approach for low power WSNs.

2.3.2 Requirement Analysis

The broad field of possible applications and the manifold approaches for WuR architecture described in the previous sections illustrate versatile requirements. In [DEO09], benefits and challenges of WuRs are discussed from functional point of view with focus on communication protocol aspects. Furthermore, a survey of state-of-the-art in research and off-the-shelf products is presented. The only WuR that is commercially available on the market is AS3933 from Austria Microsystems [35]. It consumes only 4.1 µW of power, but it is designed for operation with low frequency carrier from 15 kHz to 150 kHz and for three-dimensional near-field communication with short-range links. So typically it is applied in active RFID tags well in opposite to the anticipated utilization for WSN nodes, where a radio link distance of at least some meters

Related Work

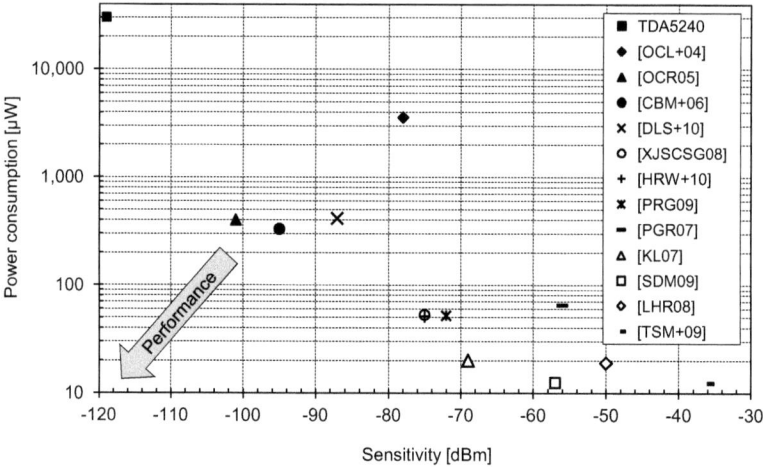

Figure 2.19: Tradeoff between power consumption and receive sensitivity of dedicated wake-up receiver designs from literature and low power main receivers

and extra margin for sufficient reliability is required.

The following items summarize and conclude most important requirements as well as their tradeoffs for hardware design of WuRs:

- Ultra-low power consumption: The power demand of modern microcontrollers in sleep mode with enabled realtime clock is in the range of only a few microwatts. In many ultra-low power sensor node applications, this power consumption becomes significant in case of low network traffic and seldom measurement tasks, where the WSN node stays in standby most of the time. As a consequence, the goal is to have a WuR whose power demand is in the same order, so that node lifetime then would not be limited predominantly because of the WuR's energy consumption. Thus, power consumption of state-of-the-art wake-up receivers presented in the previous section has to be reduced further. It should be well below 5 µW to avoid a system bottleneck in energy consumption and allow button cell powered wireless sensor nodes with lifetime of several years. Due to this very limited power budget, power hungry RF circuits cannot be applied and this has strong impact to WuR architecture. Additionally, different power states such as active RF listening mode, standby mode, or power-down mode allow flexible adjustments according to particular application demands.

- High receive sensitivity: When compared to off-the-shelf main receivers, the wake-up receiver's main penalty is low receive sensitivity and hence, a reduced wake-up range mainly as a result of power consumption tradeoffs. Solutions for compensation of this system bottleneck from [SWP10] are denser deployment of wireless sensor nodes or scaled-up transmit power as long as sufficient energy reserve is available and if regulation standards are met. Additional amount of energy and higher node density would raise system cost significantly, so enhancement of sensitivity is the main direction of impact. Very simple and strait forward implementations for WuRs such as an antenna that is directly connected to a logic gate input will work with nearly zero demand of standby power, but only at lots of RF signal strength. The penalty is unsuitable low sensitivity. An analysis of the RF-link budget for applications with in-car communication scenario from [DRK10] shows path loss of up to 68 dB at 868 MHz. Similar measurements from [DR11] for wireless communication within an aircraft result in path loss of up to 61 dB and a fading depth of up to 17 dB. So further margin is essential to tolerate fading effects. High immunity and robustness against interference is required to minimize probability for noise-caused false wake-up events. Common RF modulation schemes such as ASK/OOK or FSK, usual data rates and first of all low bandwidth occupancy are important for practical application. The target is to achieve a WuR sensitivity of around -70 dBm for adequate applicability.

- Low latency: Low delay in wireless communication that is mainly caused by the node wake-up process from low power standby mode is the major intention of WuRs. In contrast to time division multiple access (TDMA) based wake-up methods with duty-cycled main receivers, always listening WuRs are capable to handle realtime communication requirements. However, the maximum length of wake-up packets is limited. Tolerable latencies for devices with human interaction are typically in the range of a few milliseconds. For the presented automotive and aeronautic application scenarios, a maximum latency below 10 ms is required.

- Addressability: In order to generate adequate wake-up interrupts for specific sensor nodes, the WuR's minimal functionality is addressability. This includes wake-up of dedicated nodes via uni-cast addressing as well as multi-cast or broadcast addresses for concurrent wake-up of node clusters or even all network nodes within range of coverage. Addressing is ensured with the help of a dedicated wake-up packet that is customized for the characteristics of the specific WuR architecture. It includes at least a preamble and some kind of address information, whereupon the preamble helps to prepare and synchronize the WuR for address decoding. Long addresses support large network size, but increases latency because of elongated wake-up packet length.

- False wake-up rate (FWR): Unwanted wake-up events can occur due to the permanent impact of noise or interference that can trigger a wake-up event. It is mandatory that probability for unnecessary false wake-ups is sufficiently low. A false wake-up would force the main receiver to be enabled to listen to the radio channel for an incoming data packet that is never received. So this activity causes additional energy consumption. As long as wasted amount of energy due to false alarms does not raise overall consumption significantly, extra energy demand is negligible. Consequently, an acceptable FWR has to be very low in the range of just a few false alarms per hour. On the other side, probability for missed wake-up detection must be low to avoid overhearing of wake-up packets. This circumstance determines the sensitivity boundary of the WuR at low receive signal strength as well as robustness in case of wrong address decoding due to perturbing interferers.

- Flexibility: The requirements for WuRs are manifold. An almost optimal match of performance capabilities to the particular application demand or application state is essential for efficient operation of a sensor network. So flexible configurability can help to adjust inherent tradeoffs and therefore achieve a best performance compromise within available abilities. Typical such tradeoffs are data rate/wake-up latency, receive sensitivity, power consumption, and immunity against interference. Moreover, versatile configuration options for clock generation, signal conditioning, calibration, decision thresholds, RF channel selection, and various modes of operation boost coverage of a broad application field.

- Low system cost: High grade of integration density saves space but also cost. So a single chip solution with almost no external components is the ultimate design goal. Standard CMOS technology offers the potential for high density SoC integration together with digital circuits. The WuR then would be part of a sensor node ASIC that includes main transceiver, microcontroller and also sensor devices or actuators. The presented state-of-the-art WuR designs occupy a semiconductor area well below $1\,\text{mm}^2$ even with an integrated matching network. In some cases, SiP integration of BAW filters for RF channel selection was demonstrated, but normally, frequency references such as quartz crystals remain off-chip. When compared to main receivers, the space and cost that is necessary for additive integration of WuR functionality is marginal, so the benefit of the extra WuR feature prevails.

However, when compared to conventional main receivers, there always has to be a compromise in performance because reduction of power consumption is not for free. Also for an ideal receiver architecture and perfect implementation, the remaining tradeoffs are power consumption, receive sensitivity, and data rate and thus latency. At the end, these three dimensions represent the theoretical limit in wireless communications. Reduction of current consumption with advances in CMOS technology progresses much slower in analog domain than in digital circuits. So the main direction of impact for WuR design are reduction of bandwidth and data rate as long as latency requirements are met. This way, the tradeoff between latency and power consumption is utilized to save power via reduction of information transfer rate to a minimum necessary. So the anyway low receive sensitivity of WuRs is not degraded collaterally.

3 Solution Concept

To cope with the compromise between power consumption and responsiveness of common transceivers, a wake-up receiver (WuR) based architecture is proposed. Then there is no need for power-saving but complex scheduling protocols with long hop-by-hop latencies, or a simple but power-hungry idle listening receive mode. In addition to the main receiver, the ultra-low power WuR is used for RF channel listening and wake-up detection, so the main receiver can be shut down until a wake-up event occurs and the standard data communication phase starts. Apart from an auxiliary wake-up preamble, this approach offers the benefit of no further delays during communication. As discussed in the previous sections, there are many competitive requirements for an appropriate solution.

This chapter presents the design of a complete add-on WuR solution for wireless sensor nodes. Beside actual WuR functionality, this includes off-chip components for RF filtering and antenna matching as well as a serial seripheral interface (SPI) for effective microcontroller communication. Development and determination process for the proposed WuR design are described. The motivation for architecture and design decisions are given, discussed and compared to existing realizations.

3.1 Design Process

Design process and design strategy have major impact to development progress. In order to come to the solution of interest with preferably optimum performance and coevally at low risk for failure and with a minimum of effort, a design process that is customized for the specific development is mandatory. Ideally, the available resources are spent in a most effective way to save unnecessary cost and minimize development time since predominantly parts of the design flow are executed manually.

Figure 3.1 illustrates the design flow that is chosen for development of the WuR ASIC. Principally, the main design approach is top-down, but design is done on different levels of abstraction more or less in parallel because of interdependencies. The three main abstraction levels are system level with description of system behavior and functionality, component level with system decomposition into detailed subsystems and components, and finally, transistor or gate level where the complete design is realized with CMOS devices. The tasks of different abstraction levels may be closely related to each other, so co-design across the three levels is the usual case. It is most likely that the solution of a first stage cannot fulfill all requirements. Therefore, an iterative design process is necessary to firstly identify difficulties, and secondly get rid of them. An iteration can cover a single or multiple tasks and may be very fast on system level or take much effort on transistor level. The eye is kept on optimization of the whole wireless system with inclusion of environment, because a global optimum is the goal and the sum of local optimization criteria does not automatically lead to an overall optimum. Otherwise, resulting bottlenecks on system level can limit overall performance.

First of all, WuR concepts and architectures are investigated. With the help of system level simulation, first parameters, critical points and difficulties are identified. So study of feasibility is possible and this allows evaluation of theoretical limitations of the system performance. These limiting factors define physical boundaries for performance and can be used later on as reference for quality of the actual design. This way, risks and performance limiting key components are identified on an abstract system level. In parallel, rough circuit design, layout and evaluation of essential and critical parts on component level and transistor level yield parameters of quite high accuracy for estimation and more detailed specification of WuR architecture on system level. Via this feedback from lower abstraction levels, the optimum structure and partitioning of subsystems of the final solution is figured out. It takes typically some iterations to distribute resources (e.g. power consumption) almost balanced to subsystems and components in order to achieve overall performance optimum as well as prove of concept and feasibility. Once the system architecture is fixed, detailed investigation and design is done on component level, whereat special attention is put to the most challenging parts to identify upcoming difficulties and risk as soon as possible and readjust specification if necessary. Most of the low level circuit design and layout is started as recently as the risk for major changes in design is cleared. Then the remaining short redesign cycles only concern iteration steps at transistor level that have minor impact to the upper levels of abstraction. After layout verification, the preliminary values estimated for system level simulation during specification phase are extracted then from final layout for a validation check. So it is possible to prove the performance via system simulation with accurate component parameters that include layout-extracted parasitic circuit elements at least of selected key components. Not till then, fabrication process for the first test

Solution Concept

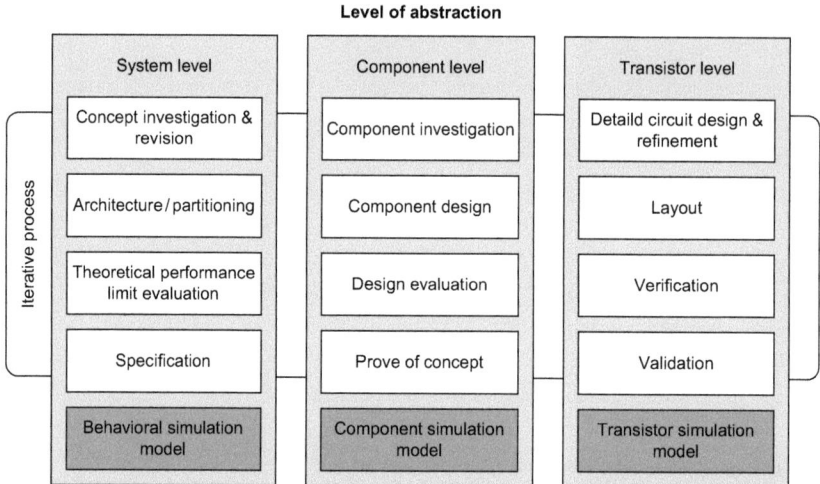

Figure 3.1: Illustration of iterative design process flow on different levels of abstraction for development of WuR ASIC

chip is started. With the help of measurements on physical prototypes, quality and accuracy of simulation models is proven. Major deviation from expected results that are larger than variation of fabrication process are analyzed and generate additional knowledge and feedback for improved (re)design of a second generation of test chips, if necessary.

In [GAGS09], Daniel Gajski proposes an approach for an automated design flow for digital domain. He starts from an executable specification and describes possibilities how to get to an ASIC layout almost without manual interaction. The idea of simulation models with different levels of abstraction is beneficial also for an analog and manual design flow.

Granularity of simulation models always defines a certain compromise between accuracy of simulation results and execution time. For example, gate level simulation is very accurate, but execution time of typical system-wide simulations that include environmental behavior would easily exceed product development time only for a single run. So different simulation models for different levels of abstraction for the same component can offer a flexible compromise for fast simulation at sufficient accuracy. The four model hierarchies used for WuR design are outlined in the following list and match to the three levels of figure 3.1 with the exception of an additional split of the low level transistor model:

- Behavioral model: This high level functional model is comparable with Gajski's model of computation (MOC) and describes pure functional behavior of the total system. The model has very sparse knowledge about underlying structure and relies predominantly on estimated parameters, but executes very fast and therefore allows short revision cycles. The simulation environment for behavioral modeling of the WuR is MATLAB. Iterative model refinement and upgraded parameters successively gain accuracy of simulation results after each step.

- Component model: The aim of this model type is to represent a component or subsystem in all its necessary details but still with mathematical and/or behavioral description. Now timing and implementation structure as well as tolerance or variations are modeled to represent the underlying circuit implementation almost closely. The model typically includes parameters such as gain, input impedance, frequency response, etc. as well as their variances and interdependencies. It again is completed with estimated influencing factors from anticipated circuit design. This model is comparable with Gajski's transaction level model (TLM) for digital domain. The huge benefits of the model type are accurate simulation results and concurrently low execution time. Such kind of models may be implemented in SystemC-AMS, VHDL-AMS [AJK05], or SpectreHDL that allow for mixed-signal simulation via their analog extension and can support modeling of other physical domains including thermal or mechanical representation. With the help of this detailed component model, system behavior of architecture and algorithms are proven for correctness. A reasonable range for possible system configuration options is identified and system performance is further optimized. Extraction of important system parameters and particularly feedback is required for refinement of system level simulation.

- Transistor level model: This simulation model consists of the final circuit design and is based on highly accurate and complex models of CMOS devices available in the specific semiconductor technology. It is similar to cycle accurate models (CAM) from digital domain. Since effects of layout are not considered here, parasitic elements at critical nodes of layout can be estimated in a first step and back-annotated from layout in a second iterative step. Circuit simulation at transistor level is possible only for limited schematic size because of the long execution time.

- Layout extracted model: Highest simulation accuracy is achieved via inclusion of parasitic elements from chip layout. These are mainly distributed series resistors of wiring and distributed capacitors that couple across signal lines. Extraction of these elements can slow down execution time enormously already for small circuits because of thousands of additional circuit elements. So typically these layout extracted models are used only to prove the final performance match to the upper transistor level model and refine estimated parasitics. Simulation of a total component with layout extracted model would

Solution Concept

take unsuitable long time or even is impossible because the solver is not able to find a solution with acceptable error due to the high complexity.

Not each component or subsystem is modeled on each level of abstraction. Detailed simulation at several levels is necessary only for key components with major impact to system performance. The main benefit is the combination of different models to a powerful mixed-model system simulation with interchangeable levels of abstraction. This flexibility accelerates development process due to fast adaption of the accuracy versus execution time tradeoff according to the actual need.

3.2 System Architecture

In order to cope with the problem of network responsiveness outlined in chapter 2 and simultaneously meet the requirements for low power consumption, the proposed approach is a highly optimized receiver with ultra-low power consumption that can stay in RF channel listening mode permanently, while consuming only a few microwatts. Therefore its own-power consumption must be lower by orders of magnitude when compared to state-of-the-art main transceivers and it is anticipated to be around $3\,\mu W$. With the help of this dedicated wake-up receiver (WuR), the main receiver (RX) can be shut down and the WuR detects packet arrival and notifies the main receiver, if a wake-up condition and a node address match are detected. Via an additional wake-up preamble in the header of the transmitted data packet, it is possible to operate the main transceiver very similar such as in the permanently powered case, but concurrently at very low power consumption. Wake-up latency for notification of packet arrival must be lower than $10\,\mathrm{ms}$ to ensure real-time capability. As a consequence, there is no need for synchronization of node clusters and thus allows for simple and straight forward MAC protocol design.

Since nothing is for free, the main compromise for this highly specialized WuR when compared to main receivers are constraints regarding RF modulation technique and significantly reduced receive sensitivity.

The proposed approach for an ultra-low power WuR concept is to eliminate all power-hungry signal processing in RF domain and combine a simplified passive RF frontend with enhanced signal processing in baseband. The intention is to compensate performance penalties of the analog RF frontend in low frequency (LF) domain, where signal processing has the potential for much reduced power demand. The proposed WuR design makes use of on-off keying (OOK) modulated carrier frequency together with a simple and "old-fashioned" RF envelope detector demodulation technique for direct down-conversion. Compared to the common superheterodyne architecture of normal main receivers, this architecture is preferred over more complex schemes to get rid of power-consuming circuits for low noise amplification and for generation of local oscillator frequency for mixing, which would easily exceed the very limited power budget by orders of magnitude. Without active signal processing in RF domain, most power can be saved, but this results in significantly reduced sensitivity when compared to off-the-shelf receivers that are based on a superheterodyne principle. The consequence would be a clear bottleneck for radio link budget of the WSN system, so enhancement of WuR's sensitivity is the major intention. This can be achieved via specific and optimized signal processing in baseband domain and via reduction of signal bandwidth to the possible minimum that in the end defines maximum tolerable latency.

Selection of the radio frequency band for a wireless system is important and very dependent from application. The WuR is designed for the 868 MHz ISM band because of excellent wave propagation especially for indoor scenarios with low probability for line of sight. The comparatively low usable bandwidth is no limiting factor because of the anyway little communication traffic in low-power sensor networks. Typical antenna size for a 868 MHz monopole is well below 10 cm and thus acceptable for most applications.

Figure 3.2 illustrates a simplified block diagram of the proposed WuR architecture. An off-chip SAW filter selects the desired radio channel for reception of wake-up packets first. Without application of a power-hungry low-noise preamplifier, signal processing in RF domain is ensured only by nonlinear envelope detection that converts OOK signals to baseband. The small signal amplitudes after direct down-conversion are amplified by a low-noise gain stage and post-processed via a transmit signal-matched processing unit in order to reduce bandwidth of system noise. This way, the limited sensitivity of the ultra-low power frontend is enhanced enormously in baseband domain. Receive address check and finally wake-up notification is ensured inherently by the signal processing unit. In contrast to many WuR related implementations

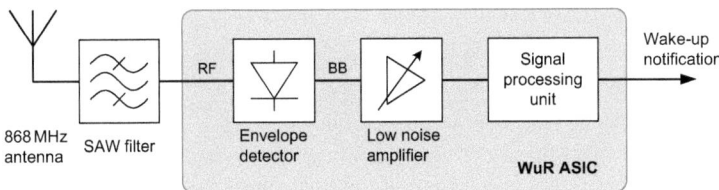

Figure 3.2: Simplified block diagram of proposed ultra-low power wake-up receiver: Approach with passive RF frontend, low noise baseband amplifier and baseband signal processing unit
RF: radio frequency, BB: baseband

Solution Concept

from section 2.3, an architecture with multiple wake-up stages is avoided. From global point of view, such a heterogeneous wake-up scheme inserts a bottleneck somewhere into the system. For example, a simple RF power detector can be used to trigger a powerful but also power-hungry address decoding stage as soon as a certain signal strength is detected. But there is the problem of low sensitivity and overhearing of weak radio signals that do not activate address processing. On the other hand, noise and interference can trigger the address decoding unit with high probability and thus power consumption would increase dramatically.

Due to the restricted power budget, a well balanced tradeoff between performance and power consumption for all WuR building blocks is mandatory. The block diagram in figure 3.3 gives a more detailed overview about the proposed architecture. Antenna signal is filtered first to suppress interference from adjacent radio channels and fed to an external power matching network that transforms impedance for maximum sensitivity. It cannot be realized on-chip because of the high quality requirement for the inductors. The following envelope detector is implemented with CMOS devices only and converts radio signal to baseband domain with high frequency-conversion efficiency. The resulting signal-to-noise ratio (SNR) after down-conversion is comparatively low, so a low noise amplifier has to avoid further degradation of noise figure. A low-pass filter with adjustable cutoff frequency and a programmable gain amplifier (PGA) ensure signal conditioning and guarantee anti-aliasing of subsequent sampling circuits. The key component is an ultra-low power correlation unit realized in analog and mixed-signal topology. It operates with digital spreading code sequences and correlates them with the incoming baseband signal to detect their amplitude within the mixture of noise and received signals. This way, high coding gain is exploited before a slicer decides if a wake-up interrupt is generated when the correlator output exceeds a configurable threshold. Alternatively, an ADC samples the analog baseband signal and puts out a serial data stream for external signal processing in digital domain to offer additional flexibility. In order to provide a full-fledged companion chip solution for common WSN nodes, configuration registers, serial seripheral interface (SPI) for microcontroller communication as well as a power management unit are added. It covers a voltage regulation part including voltage reference, a current reference for bias current generation and a clock supply unit with phased locked loop that generates a configurable system clock from a single low-frequency input. An auxiliary test interface with analog and digital outputs supports test, ASIC characterization, and debugging.

When compared to state-of-the-art, the main innovation is the additional mixed-signal correlation unit that gains SNR, receive sensitivity and simultaneously assures WuR address decoding.

Due to the compromise for low power consumption, WuRs usually have simplified RF frontend architecture such as envelope detection principle for OOK modulated carrier signals. This fact results in significantly reduced sensitivity when compared to superheterodyne receivers.

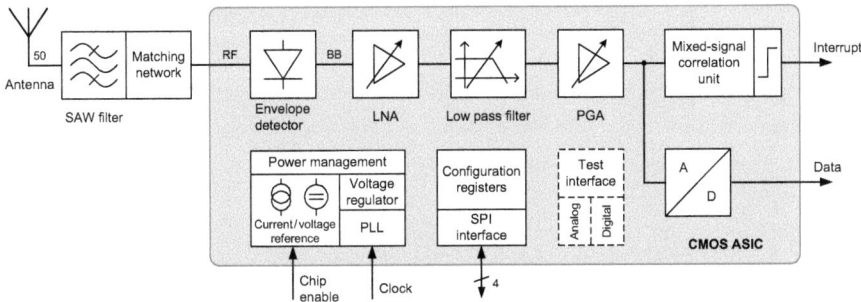

Figure 3.3: Block diagram of proposed ultra-low power wake-up receiver with passive RF frontend and correlator enhanced sensitivity
PGA: programmable gain amplifier, PLL: phased locked loop

Anyhow, to achieve a maximum of WuR sensitivity at a minimum of power consumption, the main impacts as well as their theoretical limitations for the proposed detector concept are summarized in the following list:

- RF input power matching: Since the output amplitude of a nonlinear element for RF envelope detection follows a square-law of input amplitude in a first approximation, a low-loss detector input stage together with a high-quality matching network ensure a high RF voltage transformation ratio. Therefore, detector input signal and consequently sensitivity is maximized for minimized loss in RF signal path.

- Detector sensitivity: The most nonlinear device characteristic in a standard CMOS process is the exponential function of a diode's current also known as Shockley equation

$$I_D = I_S \left(e^{\frac{V_D}{nV_T}} - 1 \right), \tag{3.1}$$

whereat V_D and I_S are the diode's forward voltage and saturation current, n is the ideality factor and $V_T = \frac{kT}{q}$ is the thermal voltage. The characteristic of Schottky diodes for RF detection is equally. Only typical values for I_S are much higher than that of silicon diodes and range up to microamperes. The forward characteristic of a MOS transistor in weak inversion region is very similar to that of a bipolar transistor and can also be modeled by a slightly modified Shockley equation whereat the essential nonlinear term $e^{\frac{V_D}{nV_T}}$ does not change in both cases. Consequently, all mentioned devices are possible good candidates for efficient detector devices with exponential characteristic.

- Noise: A further performance limiting factor is noise, which is mainly generated by the

Solution Concept

DC-biased envelope detector and the first gain stage of the baseband amplifier. Beside the CMOS technology- and transistor size dependent flicker noise at low frequencies, the always remaining thermal noise floor is the lower physical boundary. This is at least a consequence of the noisy differential resistance of the detector and/or first gain stage. Its value for a detector with exponential diode characteristic is given by the derivative of rewritten equation 3.1

$$R_N = \frac{\partial V_D}{\partial I_D} = \frac{nV_T}{I_D + I_S}. \tag{3.2}$$

One can see that the DC bias current I_D and thus power consumption is responsible at most for the equivalent noise resistance R_N. So power consumption finally determines the level of spectral noise floor in baseband domain.

- Correlation length: A spread spectrum technique is utilized to enhance receive sensitivity by means of correlation over long periods up to milliseconds range. The result is high coding gain of SNR at the correlator output due to reduction of the equivalent noise bandwidth. Generation of false wake-up events is caused mainly by the remaining noise. So the signal-to-noise ratio before decision-making has to be at least around 10 dB to guarantee sufficiently low probability for power-wasting false wake-up interrupts.

Figure 3.4 illustrates the principle of decision-making based on the proposed mixed-signal correlation approach. Thanks to time-continuous operation, the output of the correlation unit reflects the actual signal strength of the desired radio signal pattern that is additionally superposed with residual noise. If the received code sequence does not match to the WuR's pattern configuration, the correlator's output does not respond to the RF input signal. Hence, the digital code sequence acts as address information and the correlator inherently takes care of address detection. So the configurable wake-up pattern can be used for addressing specific wireless nodes. A wake-up interrupt $\overline{\text{IRQ}}$ is triggered as soon as the adjustable wake-up detection threshold is exceeded and the interrupt output keeps asserted as long as conditions are met. Therefore, the connected microcontroller is free to generate and handle interrupt requests on either static level or one of both edges.

Despite of the outlined constraints for the proposed ultra-low power WuR concept, the main innovation of this work is the combination of a simplified and power-saving RF detector frontend together with a low-power correlation technique in analog baseband domain. This architecture in conjunction with an ASIC implementation close to the physical limitations has high potential to achieve receive sensitivity values with relevance for practical application.

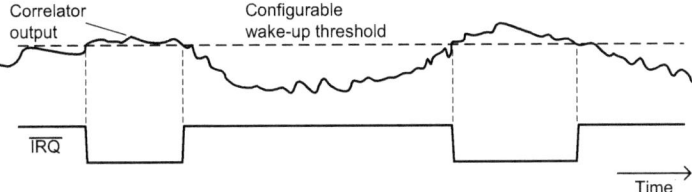

Figure 3.4: Illustration of wake-up interrupt generation from analog output signal of correlation unit

3.3 Strategies for Power Saving

Ultra-low power consumption is one of the main goals for WuR development. So careful deployment and distribution of the available power budget are mandatory. The intention is to balance power consumption between analog radio frontend and baseband signal processing unit at the backend in order to avoid a bottleneck and put the focus on maximum sensitivity enhancement of the overall WuR system. Strategies for reduction of power consumption are:

- Passive radio frontend: Avoidance of inherent power-hungry RF circuits and subsystems lead to elimination low noise preamplifiers and also mixers, because generation of necessary local oscillator frequency is very expensive. So active signal processing is no option. Therefore, an architecture only with passive impedance transformation and nonlinear signal detection remains.

- Power modes: Extensive power management with different configuration options and operation modes allows a balanced tradeoff between performance requirements versus power consumption according to the actual demand. Main power states for the proposed WuR design are an active listening mode with full functionality, a power-save mode with content retention of configuration registers for fast resume of operation as well as a power-down mode with minimized consumption. Furthermore, analog components itself can have power modes. On the one hand, voltage regulators may operate in a high-performance mode for low noise and high speed, or contrariwise in a low-power mode with reduce bandwidth, or it even can disable the output in shutdown mode.

- Duty-cycling: Also the WuR itself can be shutdown periodically for further energy saving. Because of its anyway low own consumption, extra low on-off ratios are not necessary and so node synchronization is easier. Duty-cycled operation of subsystems is another possibility to save power by means of discontinuous evaluation of analog and digital results with low bandwidth. This is ensured via power and/or clock gating.

- Ultra-low power always-on blocks: However, there always exist a few components that have to stay powered most of the time. These are typically voltage references, voltage regulators, supply supervisory circuits, oscillators or low-power wake-up timers. Since they can dominate power consumption in standby mode and may degrade battery lifetime significantly, special focus is put on development of ultra-low power solutions.

- Low supply voltage: The conventional way of supply voltage reduction for digital circuits according to speed requirements can be extended also to analog domain. Low voltage analog design in subthreshold region of MOS transistors brings additional effort [Vit03]. But if the analog supply voltage is pushed down, a common supply for digital and analog circuitry is beneficial due to elimination of level shifters and at least one additional voltage regulator. The design of digital full-custom cells and power-optimized layout for specific application within key components can save lots of static leakage current especially at elevated temperature, and also minimize dynamic consumption via reduction of wiring-capacitance.

Power Supply Unit

WSNs that are designed for supply with small batteries and for expected lifetimes of several years typically have a mean power consumption of a few microwatts. While digital circuitry in shut down mode normally has very low leakage current, the node's power supply unit stays powered all the time and has to deliver the DC output voltage as well as manage and supervise the battery or even a combination of energy harvester and energy storage device. Therefore, its own consumption should be well below the mean load current, otherwise it would shorten node lifetime significantly. Off-the-shelf products for inductor-based DC/DC conversion such as XC9226 from Torex [31] consume at least 15 µA in standby mode. Also linear voltage regulators typically have a minimum quiescent current of 0.8 µA (XC6215 from Torex [31]). So the consequence is the requirement for high power-efficiency especially in case of light load condition. Further enhancement of efficiency is possible by means of advanced DC/DC architectures that address this special issue.

The proposed DC/DC converter concept from figure 3.5(a) shows an inductor-based switched mode converter with two independent input channels and two independent output channels. Since the converter core supports both buck and boost operation with continuous changeover and concurrently at high power efficiency, it is predestinated for application within self-sufficient supply units. Via channel $IN1$, the converter delivers power from an energy harvester to the wireless node by means of channel $OUT1$. Excessive energy is stored in the energy buffer using

Solution Concept

the second output $OUT2$. In case that harvested power is insufficient, energy is taken from the storage device via $IN2$ and delivered to $OUT1$. So channel $OUT1$ is the high priority output and $IN1$ is the high priority input. This multichannel operation is ensured using time-domain multiplexed control of small energy packets. According to the actual need, these packets are transferred to the appropriate output.

Figure 3.5(b) depicts a block diagram of the proposed step-up/down converter core. The power stage consists of a matrix of 6 power switches with belonging gate drivers and supports all necessary switching states for multichannel buck-boost conversion. The power switches are controlled by a digital state machine with 4 inputs from voltage supervisory comparators and with additional 4 inputs from current comparators that detect the excess of certain current thresholds by means of a current sense amplifier. The principle of combining buck-boost mode operation with multichannel output has been recently published in [XLHK11, MRM07]. But the main challenge is an adequate switch control with smooth and fast changeover between buck and boost mode as well as minimized own-power consumption. This can be ensured by the proposed concept with a minimum number of analog always-on components in combination with an event-driven state machine that does not need continuous clock frequency. Only the voltage comparators for supervision of maximal/minimal output/input voltages stay powered all the time. The high-speed components for current measurement are not needed in idle state and therefore powered down. During the other states with anyway high energy transfer, their power consumption is comparative negligible. So the remaining ultra-low power parts are 4 voltage comparators with moderate speed requirements and a single voltage reference. This approach achieves simulated quiescent current consumption below 200 nA in standby mode with zero load current. High conversion efficiency over a large load range is achievable thanks to the current-controlled pulse-frequency-mode operation. Since an optimum power supply solution may be highly application dependent, flexible configuration options for output voltages and voltage/current thresholds via a microcontroller interface can support different types of energy harvesters and energy storage devices. This innovative approach leads to high conversion efficiency especially for light loads and supports high integration density as well, because of the single off-chip inductor for multiple I/O channels and the combined buck-boost operation mode.

Nevertheless, in some cases linear voltage regulators can provide better power efficiency than switch-mode DC/DC converters because of reduced own consumption. LDOs need weather an oscillator nor switch drivers and also the power transistor is smaller and thus has lower leakage current. Without necessary off-chip devices, cost factor is clearly an advantage.

Figure 3.6 illustrates the proposed concept for an "efficient" linear voltage regulator. In opposite to conventional LDOs with constant quiescent current, the suggested solution has an

Solution Concept

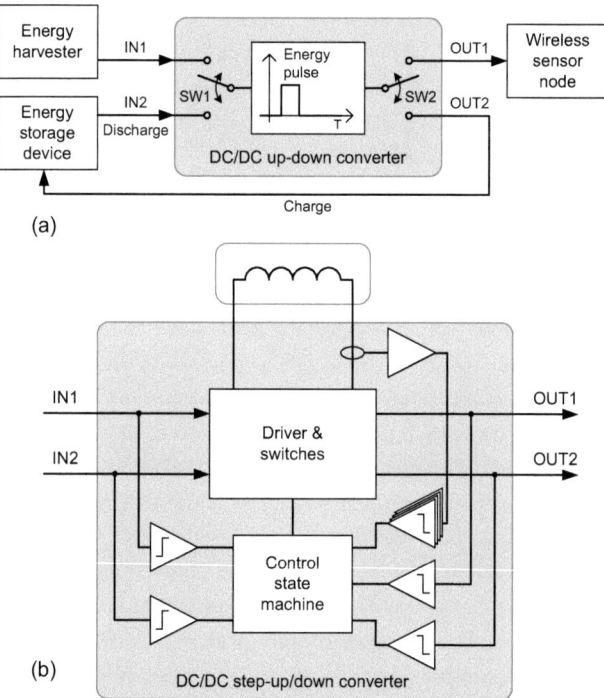

Figure 3.5: Block diagram of proposed DC/DC converter concept with high efficiency at light load condition: (a) Time domain illustration of multichannel operation and (b), architecture of inductor based switched-mode converter core

adaptive control of its own-current consumption. In order to balance the performance versus power consumption tradeoff for the linear voltage regulator, the quiescent current is instantaneously steered by the actual demand of load current. This way, the quiescent current is some 5 % of the load and this ensures a good speed versus consumption compromise both at full and also at light load condition. The disadvantage are non-constant design parameters for the control loop, so design of loop stability is critical, especially when stable the region should include resistive as well as capacitive load impedance.

One important component that is needed also for reference current generation is a voltage reference circuit. When used for low-power voltage regulation, ultra-low current consumption significantly below 100 nA is requested to avoid degradation of system power efficiency during standby mode, where the load current is typically below 1 µA. The well-known bandgap reference compensates the negative temperature coefficient (TC) of a diode's forward voltage

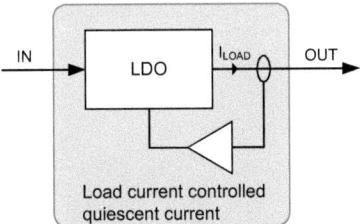

Figure 3.6: Block diagram of proposed linear voltage regulator with load dependent dynamic control of quiescent current.
LDO: low drop output voltage regulator

via addition of a certain voltage with opposite TC [Ann98]. In all implementation variants, a polysilicon resistor bridges at least some 500 mV. Therefore, it should be clearly larger than 10 MΩ in order to guarantee sufficiently low current consumption. But this is impossible because of silicon space limitations and parasitic stray capacitance, so alternative solutions are required.

Figure 3.7(c) shows the simplified architecture of the proposed voltage reference without startup circuit. The design goal is minimum current consumption and also good dynamic characteristics such as short startup time and high power supply rejection ratio. The design consist of a self-biased beta-multiplier structure (P_1, P_2, N_1, N_2, R_1) for generation of reference current I_{REF} with proportional to absolute temperature (PTAT) characteristic [dCFP05]. All MOS transistors operate in weak inversion region, so the voltage drop across the single polysilicon resistor R_1 that determines bias current is $V_{R1} = nV_T \ln\left(\frac{W_{N2}/L_{N2}}{W_{N1}/L_{N1}}\right)$, with thermal voltage $V_T = \frac{kT}{q}$, subthreshold slope factor n and transistor width W and length L. For typical MOS dimension ratios smaller than 10, the resulting voltage V_{R1} at room temperature is below 60 mV and proportional to V_T. The beta-multiplier can operate already with a few nano-amperes of reference current I_{REF} because of the comparative low PTAT voltage V_{R1}. Gate voltage V_G of the diode-connected MOS transistor N_{DIO} in figure 3.7(a) has negative TC when biased with constant current I_{DIO}. In order to design a temperature-stable reference voltage, two quantities with opposite TCs are added with proper weighting for compensation to achieve zero TC. So the negative TC of the MOS diode N_3 is canceled with the help of a serial connected stack of MOS voltage dividers from figure 3.7(b). Such dividers have a well-defined PTAT voltage characteristic across the bottom branch and it further is widely independent from drain current I_1 as long as both devices are in weak inversion [Vit09]. So PTAT voltage for equal gate width of N_{DIV1} and N_{DIV2} then becomes

$$V_2 = V_T \ln\left(1 + \frac{L_{NDIV2}}{L_{NDIV1}}\left(1 + \frac{I_2}{I_1}\right)\right). \tag{3.3}$$

Solution Concept

It is defined only by temperature voltage V_T, gate length ratio, and optionally by current ratio. Hence, the linear fraction of the MOS diode's TC can be canceled for an almost temperature independent $V_{REF} = V_{DIO} + V_{DIV}$.

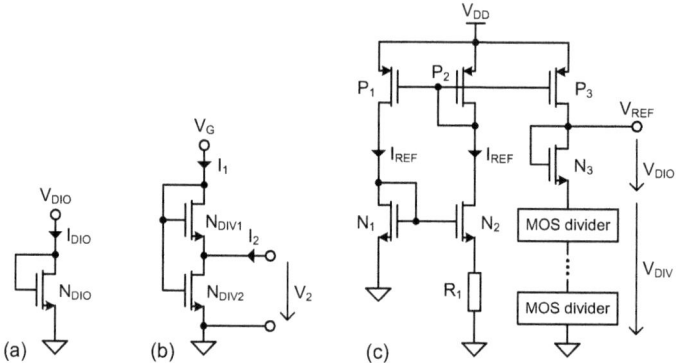

Figure 3.7: (a) Diode connected MOS, (b) MOS transistor voltage divider from [RA05, Vit09] with PTAT characteristic and (c), architecture of proposed ultra-low power voltage reference
PTAT: proportional to absolute temperature

This approach is analyzed in theory:
The equation for the MOS transistor's drain current I_D in weak-inversion region is very similar to that for bipolar transistors [RA05, Vit09].

$$I_D = 2\frac{W}{L} n_i \mu_{eff} C_{ox} V_T^2 e^{\frac{V_P}{V_T}} \left(e^{-\frac{V_S}{V_T}} - e^{-\frac{V_D}{V_T}} \right) \tag{3.4}$$

with pinch-off voltage

$$V_P = \frac{V_G - V_{th}}{n}, \tag{3.5}$$

where W and L are the MOS gate dimensions, V_G, V_D, V_S are the accordant potentials at transistor's contacts, and V_{th} represents its threshold voltage. Technology dependent parameters are intrinsic carrier concentration n_i, channel mobility μ_{eff}, and gate oxide capacitance C_{ox}. Assuming that $V_S = 0$ and $V_D \gg V_T$ in case of operation in saturation region, equation 3.4 gets

$$I_D = 2\frac{W}{L} n_i \mu_{eff} C_{ox} V_T^2 e^{\frac{V_P}{V_T}} = I_s e^{\frac{V_P}{V_T}}. \tag{3.6}$$

Temperature characteristics of mobility and carrier concentration are $\mu_{eff}(T) = \mu_{eff}(T_0) \left(\frac{T}{T_0}\right)^{-r}$ with $r = 1.4 \ldots 1.6$ from [CG00], and respectively $n_i(T) = n_i(T_0) \left(\frac{T}{T_0}\right)^{\frac{3}{2}} e^{\left(-\frac{E_g(T)}{2kT} + \frac{E_g(T_0)}{2kT_0}\right)}$ from

[CG00] with bandgap energy $E_g(T) \approx 1.206 - 2.73 \times 10^{-4}T$ for $T \geq 250\,\text{K}$, and hence $1.12\,\text{eV}$ for room temperature. So the technology specific current I_s results in

$$I_s(T) = 2\frac{W}{L}n_i(T)\mu_{eff}(T)C_{ox}V_T^2 = C\,T^{(\frac{7}{2}-r)}e^{-\frac{E_g(T)}{2kT}} \qquad (3.7)$$

with a temperature independent constant C. Rewriting equation 3.6, $V_P = V_T \ln\left(\frac{I_D}{I_s}\right)$, and the TC of the pinch-off voltage V_P is

$$\frac{\partial V_P}{\partial T} = \frac{\partial V_T}{\partial T}\ln\left(\frac{I_D}{I_s}\right) - \frac{V_T}{I_s}\frac{\partial I_s}{\partial T} \qquad (3.8)$$

when considered that I_s is a function of temperature in equation 3.7 and assumed that I_D is held constant. Hence from equation 3.7,

$$\frac{V_T}{I_s}\frac{\partial I_s}{\partial T} = \frac{V_T}{T}\left(\frac{7}{2} - r + \frac{E_g}{2kT}\right), \qquad (3.9)$$

and expression 3.8 then evaluates to

$$\frac{\partial V_P}{\partial T} = \frac{V_P}{T} - \frac{2V_T}{T} - \frac{V_{BG}}{2T} \qquad (3.10)$$

for $r = 1.5$ and with bandgap voltage $V_{BG} = E_g/q$. Finally, the temperature coefficient of the MOS diode is

$$\frac{\partial V_G}{\partial T} = n\frac{\partial V_P}{\partial T} = -\frac{1}{T}\left(\frac{nV_{BG}}{2} + 2nV_T + V_{th} - V_G\right). \qquad (3.11)$$

Cancelation of the linear TC by dint of an appropriate positive TC leads to

$$\frac{\partial V_G}{\partial T} + N\ln(1+K)\frac{\partial V_T}{\partial T} = 0, \qquad (3.12)$$

when equation 3.3 is adapted for N divider stages with gate length ratio K. Therefore, the essential condition for MOS gate dimension at zero TC is given by

$$N\ln(1+K) = 2n + \frac{\frac{nV_{BG}}{2} + V_{th} - V_G}{V_T}. \qquad (3.13)$$

Consequently, the final reference V_{REF} is the voltage sum across MOS divider stack and MOS diode and results in

$$V_{REF} = V_T N \ln(1+K) + V_G = \frac{nV_{BG}}{2} + 2nV_T + V_{th}. \qquad (3.14)$$

The reference voltage is defined by physical constants in a large extent. The only CMOS tech-

Solution Concept

nology dependent parameters are threshold voltage V_{th} and subthreshold slope factor $n = 1.19$ that has low standard deviation. So the remaining process-caused variability of the reference voltage V_{REF} is not as good as from bandgap references, but sufficient for many applications with demand for extremely low power consumption.

3.4 Radio Frequency Frontend

The radio frequency frontend has most impact to the WuR's sensitivity performance. In order to achieve high SNR in baseband domain and thus high receive sensitivity, optimized conversion efficiency of the RF detector circuit as well as suppression of noise sources are the main aims. Both must be ensured at low power consumption.

3.4.1 Envelope Detector

High down-conversion gain of envelope detectors implies a detector device with highly nonlinear characteristic. For an integrated solution in standard CMOS technology, the most nonlinear function that is usable for RF detection is the exponential characteristic of a diode or of a MOS transistor in weak inversion. This relation is typically modeled with the Shockley equation [Apa88]

$$I_D = I_S \left(e^{\frac{V_D}{nV_T}} - 1 \right). \tag{3.15}$$

A Taylor series approximation for the exponential function around bias voltage V_B leads to

$$\frac{I_D}{I_S} = -1 + e^{\frac{V_B}{nV_T}} \sum_{k=0}^{\infty} \frac{1}{k!} \left(\frac{V_D - V_B}{nV_T} \right)^k, \tag{3.16}$$

and hence, high order nonlinear terms for $k \geq 2$ are obtained. Only parts of even order are of interest because of the requirement for self-mixing down-conversion and thereof again, just the square-term brings relevant benefit. One reason are the anyway reduced coefficients for the kth order fraction due to the $\frac{1}{k!}$ law, and the main disadvantage is the rapid drop-off of conversion ratio with decreasing amplitude A of an input voltage $V_D = A\cos(\omega t)$ as a consequence of the kth order characteristic. Hence, the remaining square-law detection principle utilizes the fact that $A\cos^2(\omega t) = \frac{A}{2}(1 + \cos(2\omega t))$. The doubled frequency as well as frequency conversion products of other nonlinear terms are filtered out and the residual DC fraction $\frac{A}{2}$ allegorizes desired signal after down-conversion.

Figure 3.8 illustrates envelope detector transfer characteristic in double-logarithmic scale. The low frequency (LF) output signal magnitude is depicted versus input signal strength of an OOK modulated RF carrier. At high input power, the detector operates in linear region and thus, it demodulates the input signal such like ideal rectification. Doubled input amplitude results in doubled output amplitude in LF domain. Contrarily in small signal region, the output has square-law characteristic due to equation 3.16, so halved input magnitude quarters the LF output signal. In double-logarithmic scale, this circumstance yields to the slope of 2 dB/dB within the transfer characteristic. At the same time, it gives the main disadvantage of low conversion gain for weak RF signals. If compared to superheterodyne concepts with linear conversion gain, receive sensitivity suffers, but power consumption may be very low.

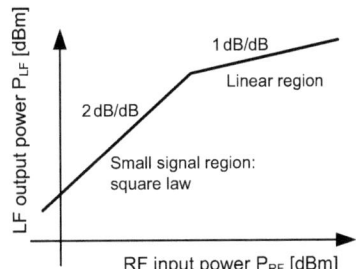

Figure 3.8: Illustration of square-law RF detection principle

Regardless if the actual detector device is a silicon diode, a Schottky diode, a bipolar transistor, or a MOS transistor in weak-inversion, as long as the nonlinear term $e^{\frac{V_D}{nV_T}}$ in its characteristic does not change, all of these devices have the same RF-to-baseband conversion gain for equal biasing. Beside the constraint for amplitude modulation, the clear advantages of the RF detection scheme are very low power consumption, possibility for on-chip integration [JO04], and the inherent large detector bandwidth. So the carrier frequency can be changed easily from one to another frequency band via adaption of the off-chip matching network while the detector itself can operate elsewhere between some hundred megahertz and a few gigahertz.

In order to exploit maximum detection sensitivity, RF input power has to be matched properly. Because of square-law detection, the benefit of increased input signal resulting from impedance matching is disproportionately high and therefore of big importance. Figure 3.9 shows the typical structure of a matching network for diode detectors that consists of two inductors. The equivalent RF impedance of the biased diode detector includes differential resistance R_D, junction capacitance C_D, and additional series resistor R_{ESR}. From equation 3.16 one can see

Solution Concept

that maximization of detector voltage V_D is the aim for high receive sensitivity. If lossless inductors are assumed, the total RF input power is transferred and dissipated at the detector. This results in maximum voltage V_{DET} and hence, good sensitivity. So the demand for voltage matching is equally with the common power matching approach. In practice, $R_D > 10\,\text{k}\Omega$ because of the low bias current and therefore, inductor L_1 must be of high quality to ensure low-loss impedance transformation. L_1 and C_D represent a resonant circuit and mainly determine operating frequency via the resonance frequency equation $2\pi f_{res} = \frac{1}{\sqrt{L_1 C_D}}$. Consequently, low input capacitance C_D is beneficial because this yields to a larger inductance L_1 and reduces loss in the unwanted resistor R_{ESR}. Inductor L_2 finally matches the input impedance to $50\,\Omega$. Furthermore, a small bandwidth of the matching network supports preselection of desired

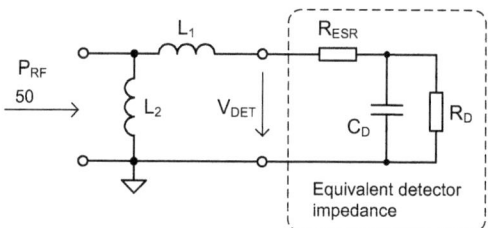

Figure 3.9: Typical matching network for RF detectors
ESR: equivalent series resistance

frequency band and filters out interference of adjacent channels. But still for most applications, an additional SAW filter with excellent out-of-band suppression of interferers such as GSM is required. This brings additional insertion loss of typically $\approx 1.5 - 2.5\,\text{dB}$.

The on-chip detector realized in [JHKT09] consumes 900 nA and has a sensitivity of -28 dBm. In principle, the proposed approach is similarly, but with deep study of interplay between detector, matching network and baseband amplifier, much better sensitivity is gettable via an optimized frontend design.

3.4.2 Noise Considerations

Another performance limiting factor is presence of noise. While down-converted noise from RF signal detection leads to negligible contribution to noise figure of total receive chain, the remaining noise sources are mainly envelope detector and first gain stage of baseband amplifier. Disturbing noise spectrum for both components is in baseband domain and is shown in figure 3.10. At low frequencies, CMOS-typical flicker noise has $1/f$ characteristic up to the flicker noise edge between approximately 10 – 100 kHz that is dependent also from transistor geometry. The thermal-caused noise floor at high frequencies is a result of the noisy differential resistance and thus a lower physical boundary. For maximal SNR, the RF detector and the low noise preamplifier must work at this thermal noise floor, so a low frequency cutoff at flicker noise edge is mandatory.

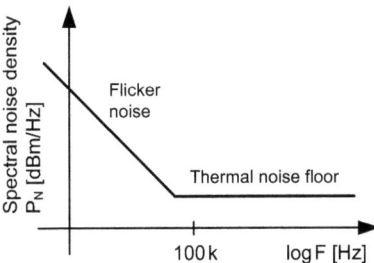

Figure 3.10: Spectral noise characteristic of MOS transistors

The no-load voltage noise V_N of a noise equivalent resistor R_N at temperature T and within bandwidth B is given by the equation for Johnson-Nyquist noise

$$V_N = 2\sqrt{kTBR_N}. \tag{3.17}$$

Insertion of equation 3.2 yields to a diode detector's RMS noise voltage

$$V_N = 2\sqrt{kTB\frac{nV_T}{I_D + I_S}} = 2kT\sqrt{\frac{nB}{q(I_D + I_S)}}. \tag{3.18}$$

So thermal noise is proportional to absolute temperature and decreases only with increased bias current I_D or high I_S such as for special Schottky diodes. Since the forward characteristic of weak-inversion MOS transistors is equally except that $I_S \approx$ pA, at least the LNA contributes significant noise or would consume excessive power. Consequently, the combination of detector and amplifier is the proposed solution out of the dilemma and drafted in figure 3.11. Transistor N_{DET} simultaneously ensures envelope detection and amplification of the demodulated baseband signal. The low pass filter R_1 and C_1 suppresses residual RF until further amplification via the PGA. This input architecture guarantees minimum noise figure at a given bias current, so I_B is chosen to determine power consumption predominantly for high SNR output. Input impedance of transistor N_{DET} is widely capacitive and has low loss, so higher voltage transformation gains receive sensitivity when compared to diode detectors. This first amplifier stage has high gain, so noise of subsequent gain stages contribute much less to overall noise figure

Solution Concept

and thus, they can operate with much reduced current consumption.

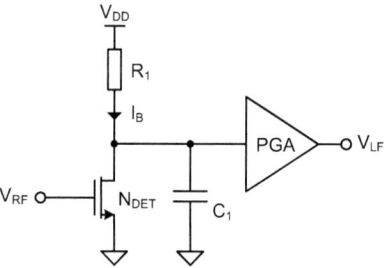

Figure 3.11: Integrated NMOS transistor detector in weak-inversion operation: The transistor N_{DET} simultaneously acts as RF detector and first gain stage of the baseband amplifier chain. Generation of bias voltage I_B is not shown. *PGA: programmable gain amplifier*

After signal processing in baseband domain, the analog output signal still contains some residual noise. So the decision threshold for wake-up event generation has to be high enough in order to guarantee sufficient low probability for false wake-ups P_{FW} during idle listening in absence of receive signal. In case of additive white Gaussian noise, probability for exceeding the wake-up threshold $V_{threshold}$ is

$$P_{FW} = \frac{1}{2}\mathrm{erfc}\left(\frac{V_{threshold}}{V_N}\right) \tag{3.19}$$

whereat V_N is the RMS quantity of noise voltage. This is exactly the same probability as for bit error ratio in common binary phase shift keying (BPSK) modulation schemes. It is also shown in figure 3.12. Here, $BER = \frac{1}{2}\mathrm{erfc}(\sqrt{SNR})$ with SNR as linear signal-to-noise power ratio. In this context, magnitude of decision threshold $V_{threshold}$ can be interpreted as signal voltage and RMS noise voltage V_N corresponds to mean noise magnitude. If false wake-up rate (FWR) should be below one false alarm per hour and evaluation period of signal output is for instance $1\,\mathrm{ms}$, then $BER < 3 \times 10^{-7}$ is required and the consequence would be high $SNR > 11\,\mathrm{dB}$.

3.5 Baseband Signal Processing

One main disadvantage of ultra-low power WuRs is their reduced sensitivity. In order to gain performance particularly for architectures with passive RF frontend, a correlation technique is utilized. This innovative concept combines simple and power-saving receiver frontends with enhanced signal processing in baseband domain to compensate their performance penalties at least partially and thus achieve sensitivities with serious relevance for practical application.

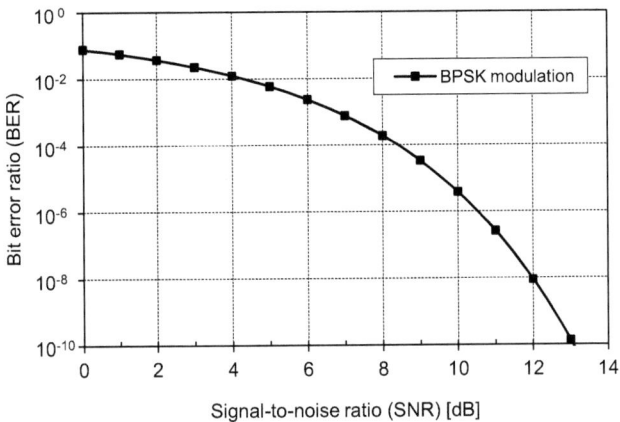

Figure 3.12: Dependency of bit error ratio (BER) from signal-to-noise ratio (SNR) for BPSK modulated signals and additive white Gaussian noise

In CDMA networks, a correlation method is used in receivers to separate spread-spectrum signals which are transmitted over a shared frequency band to establish simultaneous communication links. In opposite, the proposed correlation technique does not influence OOK modulation of the RF carrier and hence, also RF spectrum is unchanged. The approach is to apply a correlation unit to the receiver's baseband signal in order to reduce its noise bandwidth and exploit coding gain for further enhancement of SNR. This technique has high potential for low power consumption because signal processing is done in the low frequency baseband domain. In general, RF modulation scheme is not preconditioned for use with the baseband correlation unit. However, the correlator input signal has to contain the desired signal with amplitude modulation at least, and the PGA gain must be adjusted for almost linear operation without clipping to avoid performance degradation. The tradeoff for forced SNR and enhanced receive sensitivity is increased latency for transmit signal detection. But if the transmitter operates with a data rate of $R_{DAT} = 100\,\text{kbit/s}$ and if a delay T_D of up to $10\,\text{ms}$ is acceptable for adequate realtime capability, the resulting correlation length is still $R_{DAT}T_D = 1000\,\text{bit}$. In a first approximation with the assumption of statistical independency of noise, standard deviation is reduced by $\sqrt{R_{DAT}T_D} = 30\,\text{dB}$. So correlation length directly determines coding gain. However, because of the square-law from envelope detector characteristic, coding gain of $30\,\text{dB}$ from correlation would enhance receive sensitivity then only by $\sqrt[4]{R_{DAT}T_D} = 15\,\text{dB}$.

Since correlation is a liner operation, it can support an early wake-up feature for high receive signal strength. The precondition is a correlation pattern that additionally has periodic structure and consists of short code sequences that are repeated several times consecutively. The evaluated correlation result is continuously compared with the threshold for wake-up decision. Strong input signal leads to fast reach of the wake-up boundary clearly ahead of schedule, already if a single short code sequence is detected within the whole correlation pattern. Nevertheless, the potential of full coding gain is preserved. In case of low signal strength, accumulation of desired signal would last longer until the decision threshold is exceeded and a wake-up event is triggered. This way, the benefit of high field strength is utilized optimally for latency reduction and the clear tradeoff between sensitivity and responsiveness gets visible. Configurable correlation time as well as adjustable sensitivity threshold of the slicer provide flexible options according to the actual application demand.

A further advantage of correlation technique is that code sequences can be employed as address information for dedicated wireless nodes. Therefore the family of code sequences must have appropriate cross-correlation characteristic with high grade of orthogonality to enable uni-cast addressing. Otherwise codes behave like multi-cast or even broadcast addresses. Consequently, the correlation pattern should be fully configurable by the user in order to offer maximum flexibility. However, the number of code sequences with good cross-correlation is limited for a given moderate code length, so the number of addresses within a code family is restricted too.

Figure 3.13 shows circular cross-correlation function of a code family with 6 different code sequences of 63 bit length. Such maximum-length sequences are widely used for generation of pseudo-random noise patterns. While the 6 sequence combinations with a correlation function of 1 for zero bit-shift represent autocorrelation functions, the remaining 30 combinations are cross-correlation functions. Their maximum value is 0.36 and thus orthogonality is very limited already for this small set of sequences. All 6 autocorrelation functions in contrast have a minimum value of -0.016 for each bit shift unequal zero. So these sequence family is ideal for synchronization tasks, but this is of minor interest for pure WuR operation, since data reception with bit synchronization is not the primary intention. The frequency spectrum of all m-sequences in figure 3.14(a) is equally distributed and furthermore, it has minimum DC offset. So m-sequences would excellently qualify from this point of view.

The cross-correlation function in figure 3.15 depicts characteristic of the small set of 63 bit Kasami sequences. They have good orthogonality because adjacent maxima are only 0.11, but on the other hand figure 3.14(b) shows their less ideal spectral distribution with partly significant DC offset that might result in extended settling time and degrade WuR performance. Beside the clear advantage of high CDMA selectivity due to cross-correlation characteristic, the code family size of 8 is still rather small and enables only 8 different WuR addresses with uni-cast capability. Further suitable pattern families are the large set of 63 bit Kasami sequences or 63 bit Gold sequences. Both have a family size of 65 and have optimized cross-correlation function with maxima of 0.23. The ideal pattern for WuR correlation should have no offset,

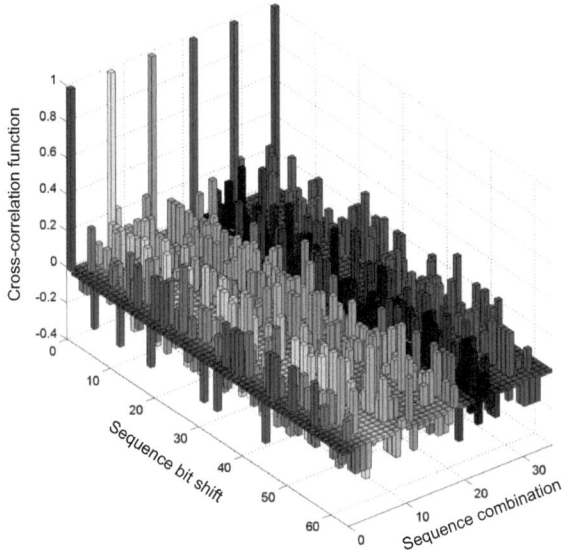

Figure 3.13: Circular cross-correlation characteristic of 63 bit pseudo-random noise patterns for correlation: The code family consists of 6 different sequences of maximum length.

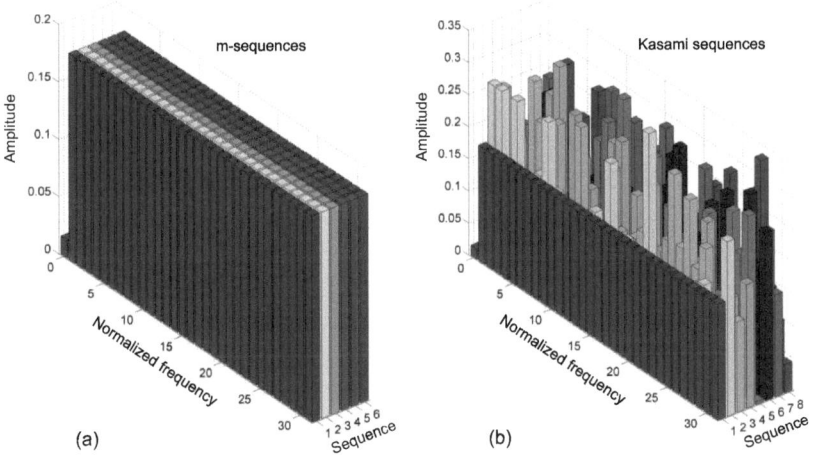

Figure 3.14: Frequency response of 63 bit code families: (a) m-sequence family and (b), small set of Kasami sequences

Solution Concept

equally distributed spectrum, almost ideal circular orthogonality and a large code set, but all these requirements are not applicable at the same time.

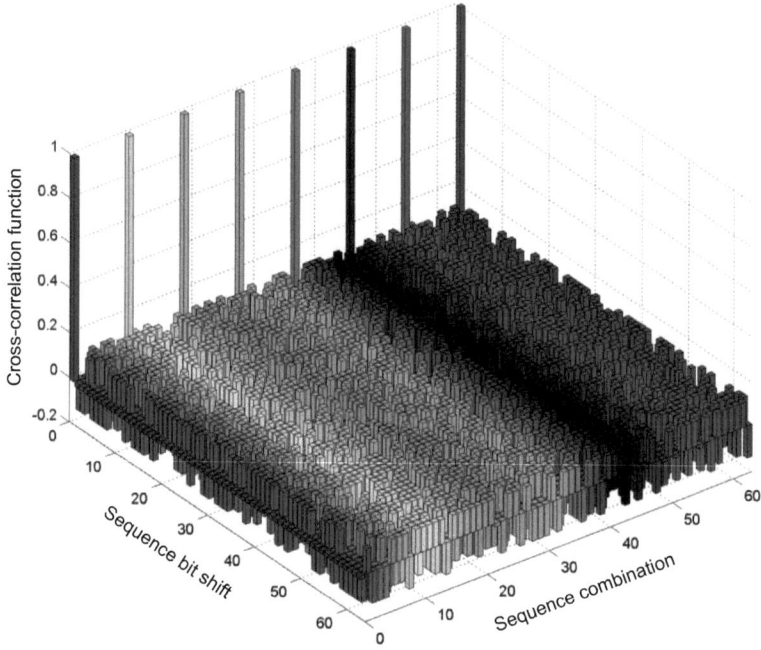

Figure 3.15: Circular cross-correlation characteristic of small set of Kasami sequences: The code family size of the 63 bit codes is 8.

However, correlation is a mathematical operation with quite much effort, so power consumption of conventional straight forward implementations in digital domain can only process with very short code sequences, or otherwise power consumption would exceed overall power budget by far. Hence, sophisticated and optimized architecture together with full custom digital design is mandatory to support preferably long code sequences. Alternatively, the preferred implementation is in analog/mixed-signal domain and provides much lower power consumption not at last due to omission of extra analog-to-digital conversion.

3.5.1 Mixed-Signal Correlation Unit

Instead of correlator implementation in digital domain, the proposed analog/mixed-signal approach benefits from a switched-capacitor (SC) principle and achieves by far lower power consumption. In conventional "serial" correlation schemes, digitized baseband signal is firstly fed though a matched filter that inverts the low-pass characteristic of the baseband amplifier chain. Then the whole sampled data sequence is shifted towards the digital pattern to build the inner product and calculate the correlation result before final output evaluation. For a correlation length over L bits, 2L samples have to be processed because of the oversampling requirement for anti-aliasing. Hence for each bit, 2L multiplications as well as a sum over 2L products have to be calculated and the complete digitized data stream must be stored for future operations. In opposite, the proposed innovative scheme operates in analog domain with "parallel" correlation stages and is presented in the block diagram of figure 3.16(a). The 1-bit digital correlation pattern is loaded into the circular shift register and filtered first in order to achieve signal-matched characteristic when compared to the desired baseband signal. Shifting and filtering only the single-bit pattern instead of the digitized baseband signal saves much power. The matched filter generates output weights of $+1$, -1, and 0 at bit transitions in the over-sampled pattern sequence, so the multiplier can be implemented simply as signed adder. Afterwards, a low pass filter accumulates pattern signal components within the desired signal and suppresses orthogonal components until a slicer with adjustable threshold detects presence of wake-up signal. Because there is no synchronization to the transmitted bit-stream, the correlator stage is implemented $N = 2L$ times in parallel where each stage is operated by a bit-shifted version of the original digital code pattern. Wired-OR connection of the decision comparators and also a single pattern shift register only with taps for each correlator stage reduce overall power consumption enormously. The complete evaluation circuitry after the low pass filters is duty-cycled without aliasing because of the reduced output bandwidth and therefore, further power is saved. The slicer threshold for wake-up interrupt generation is adjusted adequately to offer suppression of frontend noise as well as configurable amount of immunity against interference. This allows to trigger an interrupt only if a certain receive signal strength is exceeded that guarantees sufficient high probability for correct detection in spite of noise.

Figure 3.16(b) illustrates function of matched filter in signal room. The vertical vector displays transmit pattern sequence $\overrightarrow{TX}$. Due to propagation through the radio channel and bandlimited amplification within the receiver frontend, its direction gets rotated to vector $\overrightarrow{RX}_{BB}$ and a hyper-sphere of noise is added. Correlation method builds an inner product and thus extracts the baseband component that points to the same direction as reference signal vector. So the control pattern for the multiplier must also have the direction of the $\overrightarrow{RX}_{BB}$ vector. Consequently, the original $\overrightarrow{TX}$ pattern that is also available within the receiver has to be rotated via a signal-matched filter with almost equal frequency response when compared to the radio channel and the analog radio frontend. Otherwise, uncertain direction of the reference signal vector would lead to loss of signal magnitude and thus SNR. The low pass filter eliminates all

Solution Concept

L − 1 signal components out of the L-dimensional noise-sphere that are orthogonal to $\overrightarrow{RX}_{BB}$. Only the single remaining dimension adds its noise component to the desired receive signal $\overrightarrow{RX}_{BB}$ and this circumstance yields to a coding gain $G_C = \sqrt{L}$.

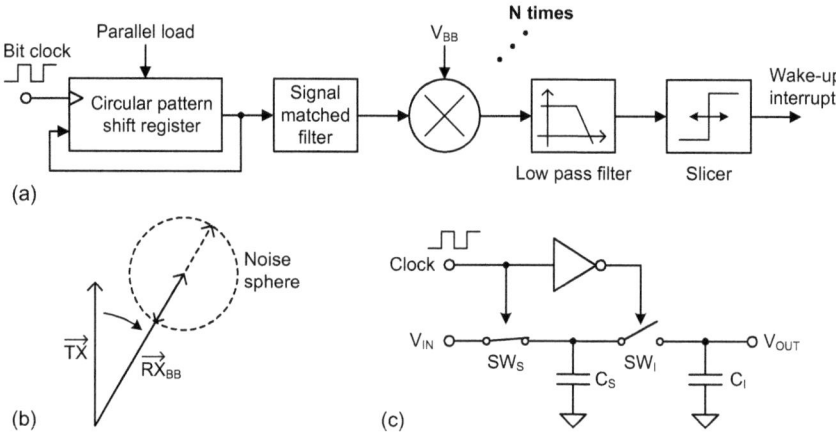

Figure 3.16: Mixed-signal correlation: (a) architecture, (b) signal room illustration of matched filter behavior and (c), low pass filter based on switched capacitor principle
TX: transmitter, RX: receiver

A time-continuous detection and evaluation scheme is realized with the help of a SC low pass filter structure from figure 3.16(c). The analog input signal is sampled via SW_S and C_S in a first clock state and its charge is then transferred via SW_I into the integration capacitor C_I during a second step before the whole cycle is repeated periodically. The normalized time constant of the filter is determined by $\tau_{LP} = \frac{C_I}{C_S}$ and can be high. So hundreds of analog samples may be accumulated and averaged via this low pass filter approach at very low power consumption.

If the correlation code sequence is retransmitted periodically several times, further correlation gain can be achieved by processing multiple identically codes consecutively by means of the circular pattern shift register. This is ensured simply via an increased time constant $\frac{C_I}{C_S}$ and without raise of complexity or power consumption of circuitry.

3.5.2 Digital Correlation

In order to support high grade of flexibility, the option for a digital implementation of the correlation unit is discussed. Beside the mixed-signal correlation approach with maximum receive sensitivity at ultra-low power consumption, digital signal processing can help to minimize degrading impact of flicker noise at low frequency more efficiently via specialized noise canceling techniques. Furthermore, the possibility of data reception including data/clock recovery by means of a conventional "serial" correlation scheme is given. The clear tradeoff is increased power consumption when compared to the analog correlation approach. So the result is a shortened correlation length for balanced distribution of power consumption between analog and digital system domains. Consequently, also the receive sensitivity suffers when compared with analog correlation approach. The essential issue is that power is deployed where the overall benefit is at most.

Evaluation and design of the enhanced signal processing unit is done in an external field programmable gate array (FPGA) in a first step, until the final solution is integrated via a power-optimized full-custom design. On-chip analog-to-digital conversion is ensured with various configuration options. This includes different types of converters, external triggered conversion with adaptive sample rate, and adjustable reference. An ADC with pulse width modulation (PWM) output provides time-continuous pulse width that is proportional to the sampled signal magnitude and hence, it has infinite amplitude resolution. Meanwhile a second A/D converter with successive approximation register (SAR) principle and low resolution directly provides digitized baseband signal. Minimizing converter resolution reduces complexity of digital logic especially for the correlation unit and saves power. If optimal adjustment of baseband signal gain is assumed, magnitude of noise takes full ADC input range because of a typically low SNR of -20 dB. If furthermore worsened noise of 1 dB is acceptable due to additional quantization noise from A/D conversion, voltage noise would increase by about 12 % and hence, a resolution of $-\text{ld}(10^{\frac{1\,\text{dB}}{20}} - 1) \approx 3$ bit is sufficient in a first approximation.

Main design aspects for the final ASIC implementation with low power consumption are an architecture with minimized number of arithmetical/logical operations, excessive options for system clocking and clock gating, and design of digital full-custom logic cells with key-functionality. Since main power is consumed by the correlation chain itself, design of a highly efficient architecture [SKP00] and asynchronous adder structures [LHSH07] can save typically up to 80 % of power when compared to conventional digital implementation.

4 Implementation

Efficient implementation of system design is challenging and requires lots of experience and knowledge in integrated circuit design. The strategy is co-design of architecture and implementation/evaluation to figure out bottlenecks and difficulties in practical realization at first. Ultra-low power design subjects are low voltage and weak-inversion operation of analog and digital circuits, design constraints of semiconductor technology, and consideration of parasitic effects such as stray capacitance or leakage currents even at elevated temperature. At the end, the intention is exploitation of the CMOS technology's full capability, but still with an eye on robustness of the design in order to cope with process variations.

The main part of this chapter introduces in functionality of the realized integrated WuR solution, and presents fundamental ASIC schematics and design decisions. Finally, selected simulation results of expected system performance are given and discussed.

4.1 Discrete Measurements

In order to prove the match of theory as well as simulation results with measurements, the proposed RF detector concept is implemented with discrete components in a first step. This approach is necessary because loss in RF path has main impact to the WuR's sensitivity and cannot be modeled accurately due to lack of proper simulation model parameters. With the help of obtained measurement results, models are updated and can be used for further design. Figure 4.1 depicts schematics of two different topologies for Schottky diode detectors. The measurement setup includes impedance matching networks for a center frequency of 868 MHz and buffer amplifiers for high impedance connection. The dedicated detector diode HSMS285 has a high saturation current of $I_S = 3\,\mu\text{A}$, so it is operated without additional bias current I_D. For simulation with the silicon diode, a bias current of $I_D = 3\,\mu\text{A}$ is chosen to ensure fair

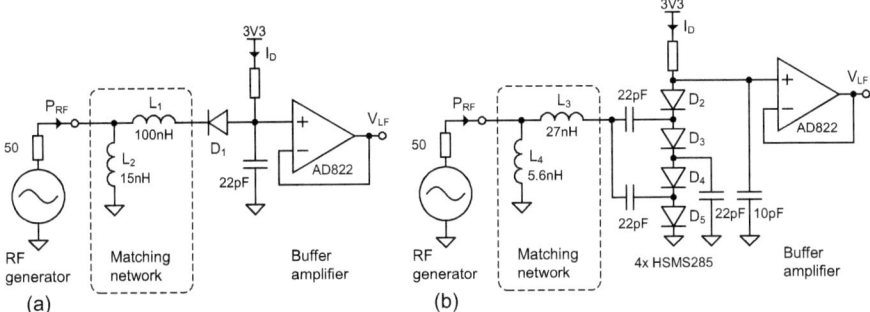

Figure 4.1: Schematic for sensitivity measurement of discrete Schottky diode detectors: (a) detector with matching network and single diode and (b), detector with Villard cascade of 4 diodes

performance comparison because of its marginal I_S.

Measurement results for diode detector sensitivities are illustrated in figure 4.2. The chart shows dependency of low frequency output voltage from RF input power in a logarithmic scale. The input signal is a 868 MHz carrier with OOK modulation at 100 kbit/s, and the result is determined via peak-to-peak voltage measurement of the first harmonic. The graph clearly shows square-law characteristic from figure 3.8 and equation 3.16 at low signal strength. Also transition to linear region at high RF input power can be obtained. The deviation of measurements from accordant simulation is very low for both detector schemes, if parasitic winding capacitance of the inductors is included. Consequently, the measurement results match to simulation models better than expected. Furthermore, the argument that silicon diodes have equal conversion efficiency when compared to Schottky diodes has been verified successfully via simulation. Hence, measurements and simulation models seem to be trustable.

When considering only conversion efficiency, both detector schemes from figure 4.1 have almost equal performance. This is a consequence of theory, at which effects of output voltage multiplication and decreased RF input impedance annihilate each other. But if thermal noise is considered too, the 4 serial connected diodes of the Villard cascade quadruplicate the equivalent noise resistance of a single diode $R_N = \frac{nV_T}{I_D + I_S}$ (equation 3.2). Thus, the SNR gets worsened by 6 dB because of doubled noise magnitude due to $V_N = 2\sqrt{kTBR_{N,equ}}$. So the envelope detector with the single diode is preferred in opposite to numerous implementations from literature, where the effect of noise is frequently ignored [JHKT09]. For practical implementation, loss in matching network relativizes this clear decision somewhat due to the drawback of a higher input impedance and therefore more delicate matching. Nevertheless, by means of high quality RF components, detector impedance of up to approximately 5 kΩ can be matched with sufficient

Implementation

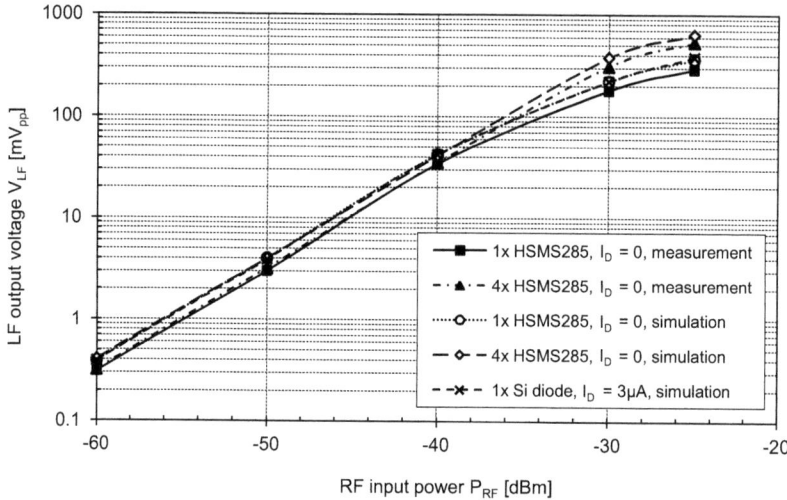

Figure 4.2: Conversion gain of different diode detectors with discrete implementation

efficiently at 868 MHz.

4.2 ASIC

The WuR ASIC is implemented in standard 130 nm CMOS technology with additional options of dual gate oxide I/O-transistors that support interfaces up to 3.6 V as well as metal-insulator-metal capacitors on top of the 7-layer metal stack for low stray capacitance to substrate. The technology further provides a high threshold voltage option for logic transistors to reduce static leakage current. The nominal supply voltage for logic cells is 1.2 V. Because of low speed requirements for the WuR design, 1.0 V supply is sufficient for digital and also for analog circuits. The main parasitic effects regarding ultra-low power design and implementation are stray capacitance and leakage. Substrate coupling, capacitance of wiring and fill structures influence bandwidth and stability of control loops. Leakage current may decrease power efficiency or accuracy and has strong temperature dependency. Also gate leakage of logic transistors can degrade performance of analog circuits with high impedance, and area restrictions limit the size of poly-silicon resistors and integrated capacitors. Ultra-low power analog ASIC design close to the technology's limitations often results in increased complexity for robust circuits with preferably design-inherent compensation of process variations and thermal effects. The intention for WuR concept implementation is exploitation of preferably full capability of CMOS

Implementation

technology for power reduction, but still with respect to robustness.
To achieve moderate cost for prototype development, a shared reticle mask set is used for ASIC fabrication. Nevertheless, more than 200 dies are expected on a single 200 mm wafer already when chip layout is placed once on mask set.

4.2.1 Block Diagram and Interfaces

Figure 4.3 presents a detailed block diagram of the realized wake-up receiver ASIC. The main building blocks are the RF envelope detector, the low noise amplifier, two AC-coupled programmable gain stages with bandwidth filter in between, and the ultra-low power mixed-signal correlation unit with adjustable decision threshold. For optional correlation in digital domain, the implementation alternatively includes an analog-to-digital conversion block with either an ADC with successive approximation principle or pulse-with modulation output. Digital signal processing is done using an external FPGA in a first step. The power management unit contains a current reference for bias generation, an ultra-low power voltage reference, and an I/O-interface with power gating options for the analog/digital 1.0 V core supply and also for the I/O voltage of up to 3.6 V. Configuration registers are controlled by means of a 4-wire SPI interface for microcontroller communication. The auxiliary test interface comprises an analog and a digital output with corresponding buffer amplifiers that provide multiplexed input channels for several internal test points. The only off-chip components are two inductors for RF power matching and a SAW filter for channel selection and interference suppression.

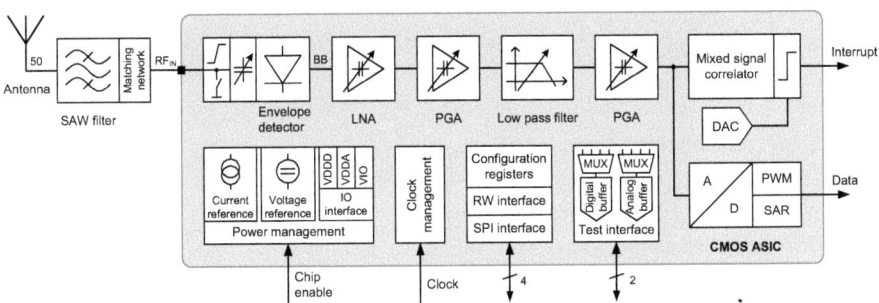

Figure 4.3: Detailed block diagram of implemented WuR ASIC with interface connections

The equivalent RF input capacitance is tunable about ±5 % to support alignment of center frequency of matching network in spite of inductance variation. Furthermore, an additional RF-switch can short the RF input signal. Then it is reflected to the antenna and the digital backend may be calibrated with the help of the residual noise signal for optimum performance.

Alternatively, the RF switch is controlled in order to limit the RF magnitude automatically and prevent the preamplifier from excessive overdrive at high signal strength. This extends the receiver's dynamic range.

Beside active mode, power management unit offers a power-save mode with register content retention and a power-down mode with chip enable pin. It can shut down the complete WuR and this results in minimum power consumption that is only caused by the remaining leakage current.

However, the ASIC is designed with extensive configuration options in order to provide high flexibility for test and evaluation of performance.

Interfaces

The single-ended RF input is AC-coupled and has antiparallel diodes to RF ground for ESD protection. No particular limitation of bandwidth allows flexible frequency band selection simply via the off-chip matching network.

The implemented 4-wire SPI slave interface is used for configuration, control and data exchange. It is compatible to SPI mode 0 and 3 and uses the signals serial clock input SCK, serial data output SDO, serial data input SDI, and low-active chip select input $\overline{CS}$ for communication. Signal-timing for read and write operations is illustrated in figure 4.4. During idle phase, SCK may be high or low. The data bits are shifted out on the falling SCK edge in both directions and are sampled on the rising SCK edge at the SDI pin. The SDO push-pull output pin is driven only during the data phase of a read command, otherwise it is tri-stated. So parallel connection of several SPI slaves is supported for interface sharing with other peripherals. Data is always coded most significant bit (MSB) first and the module is reset when $\overline{CS}$ is high.

A low level at the chip enable input EN powers down the WuR and disables its SPI. Then, the WuR switches all digital outputs to high-impedance state and all digital input drivers are disabled, so the input pins can be left floating without raise of power consumption. A rising edge at the EN input resets all configuration register content to its default value within 200 ns. By means of the SPI module, the WuR supports command based communication with direct access to configuration registers. A command consists of a 2-bit command specifier (CMD1:CMD0), a 6-bit register address (A5:A0), and 8-bit data (D7:D0). Table 4.1 shows data format of the realized read/write commands. The leading two bits of an instruction specify the type of request.

A digital input clock is required for signal postprocessing. The clock frequency is either used for sampling and analog-to-digital conversion of the baseband signal in case of digital correlation, or utilized by the mixed-signal correlation unit. So it is directly related to the transmitter's bit

Command	Bit sequence	Return value	Description
RD	[0, 0, A<5:0>, –]	D<7:0>	Read register
WR	[1, 0, A<5:0>, D<7:0>]	–	Write register

Table 4.1: Supported control commands of designed WuR ASIC

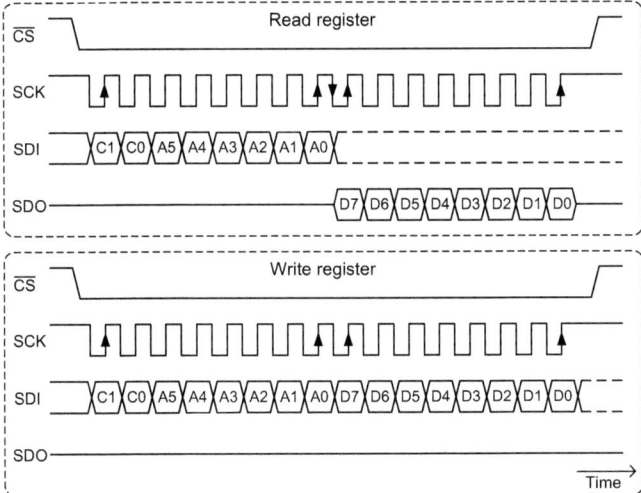

Figure 4.4: Timing diagrams for read and write operations of the implemented SPI interface

rate and hence, clock accuracy has to be better than $\frac{1}{2} \times correlation\text{-}length$ for an acceptable clock drift with phase error lower than a half bit-length.

The interrupt output pin is used to notify a microcontroller at specified events. It may be configured for push-pull or open drain output and its logic state may be active low or active high.

Supply of the WuR core is divided into an analog/RF domain and a digital/postprocessing domain with according ground connections for decoupling of disturbance. Drivers and level shifters of the I/O interface translate 1.0 V core signals to I/O voltage level of 1.2 – 3.6 V.

4.2.2 Selected Schematics

The circuit design of the WuR implementation contains 60 pages of schematics in total, already when circuit diagrams for logic cells, I/O pad cells, and ESD structures are excluded. A selected set of schematics of the most important ASIC circuits is discussed. The diagrams given in this

Implementation

section are simplified versions of the original schematics for ASIC fabrication, but represent entire functionality.

Preamplifier

Figure 4.5 shows a simplified schematic of the WuR's RF input structure. It includes two high quality inductors L_1, L_2 for low-loss RF voltage transformation together with the ASIC's mostly capacitive input impedance, the envelope detector, and the two gain stages of a low noise preamplifier. Transistor N_1 is controlled automatically to limit RF magnitude and thus detector overdrive at high signal strength to extend the WuR's dynamic range. In standard CMOS technology, p-channel metal-oxide semiconductor (PMOS) devices have a significantly reduced flicker noise coefficient when compared to NMOS. For that reason, P_1 operates in weak inversion region and acts as envelope detector. The key advantage of this principle is on the on hand the mostly low-loss capacitive input impedance of the MOS detector that allows for a high RF voltage transformation ratio compared to diode detectors. On the other hand, P_1 simultaneously acts as first gain stage of the baseband LNA. Both functionalities are achieved with the same bias current $I_{DET} = 1.0\,\mu\text{A}$. In order to ensure low thermal noise level, I_{DET} is chosen to consume the WuR's predominant power demand of $1\,\mu\text{W}$ just for P_1. In this case, the resulting noise equivalent differential resistance of P_1 is

$$R_{d,P1} = \frac{nV_T}{I_{DET}} = 30.5\,\text{k}\Omega \qquad (4.1)$$

with subthreshold slope-factor $n = 1.19$. Because bias current $I_{DET}(T) = I_{DET}(T_10)\frac{T}{T_0}$ has PTAT characteristic, temperature drift of baseband gain $G_{P1} = \frac{R_1}{R_{d,P1}} = \frac{qR_1 I_{DET}(T_0)}{nkT_0}$ is compensated for temperature stability. To enhance the gain at low frequencies in spite of a small integrated blocking capacitor $C_{block} = 400\,\text{pF}$, the signal is partly fed back via P_2 and R_3 after the RF suppression filter that consists of R_1 and C_6. Bias current of the second gain stage P_2 is only $I_2 = 160\,\text{nA}$ because of its reduced contribution to overall noise figure. The complete detector/LNA block has 30 dB gain, a bandwidth from 100 Hz to 1.1 MHz and consumes $1.25\,\mu\text{W}$.

Because of strict low-loss design in RF signal path, the matching network has comparatively low bandwidth. This helps for out-of-band interference suppression, but requires calibration of center frequency for maximum receive sensitivity. So the additional input capacitor C_1 is digitally tunable about 60 fF in order to allow adjustment of total RF input capacitance by some $\pm 5\,\%$ for compensation of L_1 tolerance. Figure 4.6(a) depicts the common approach for switching on/off a capacitor C_{SW}. Depending on transistor size, the drawback is either high stray capacitance or high on-resistance. In opposite, capacitance of the implemented tunable

Implementation

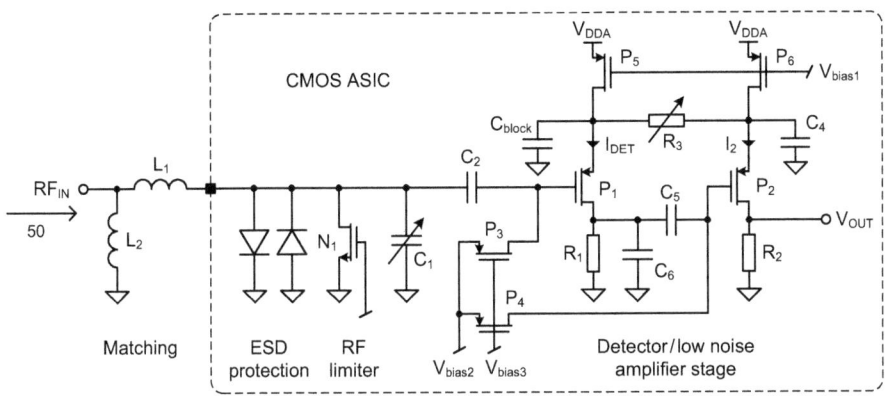

Figure 4.5: Simplified schematic of RF input structure: Transistor P_1 operates in weak inversion, ensures RF envelope detection, and simultaneously represents the first gain stage of the low noise preamplifier in baseband domain.
ESD: electrostatic discharge

MOS capacitor C_{MOS} from figure 4.6(b,c) has very low loss and it is highly dependent from gate bias voltage V_{Ctrl}. By utilizing the nonlinearity of MOS gate charge, a binary-weighted ladder of MOS capacitors is used to implement the digitally tunable capacitor C_1 from figure 4.5.

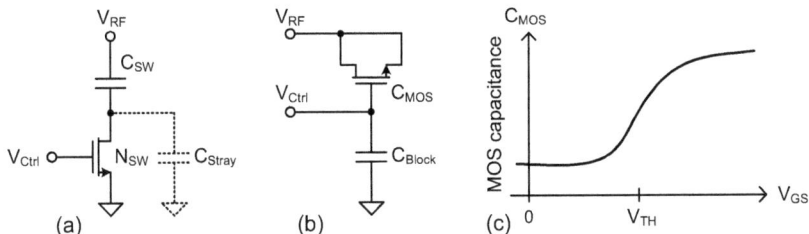

Figure 4.6: (a) Standard approach for RF capacitor switching, (b) implementation of high quality capacitor tuning with low equivalent series resistance and (c), characteristic of MOS capacitance versus gate voltage

Amplifier Chain

After low noise pre-amplification, baseband signal is conditioned via two PGAs and a second order low pass filter with adjustable bandwidth. Each of the AC-coupled PGAs in figure 4.7(a) provides configurable gain of 14 – 24 dB for almost optimal adjustment of output amplitude

Implementation

before postprocessing as well as 700 kHz bandwidth at a power consumption of 200 nW. The filter in between blocks flicker noise at the low frequency edge and prevents aliasing in the subsequent sampling circuits of the correlation unit or the ADCs. Variable sample rates are supported via digital adjustment of filter bandwidth. The simplified PGA schematic in figure 4.7(b) shows an operational amplifier based structure with adjustable feedback network for gain selection. Because of the control loop and the compensated frequency response of resistor network R_2, amplifier gain is precise and stable over temperature. The output buffer reduces output impedance and consists of N_4, N_5, and compensation capacitor C_C. In order to achieve low frequency cutoff at $f_{CU} = 40\,\text{Hz}$ for the AC-coupled input already with a small coupling capacitor $C_1 = 1.0\,\text{pF}$, transistor N_2 must have very high impedance $R_{N2} = \frac{1}{2\pi f_{CU} C_1} = 4\,\text{G}\Omega$. N_1 and N_2 operate in weak inversion. With the help of equation 3.4 and assumption of constant gate voltage V_G, impedance of N_2 is determined via $R_{N2} = \frac{\partial V_D}{\partial I_D} = \frac{V_T}{I_S + I_{Res}} \frac{W_{N1}/L_{N1}}{W_{N2}/L_{N2}}$. Consequently, the input impedance is given only by thermal voltage V_T, bias current I_{Res}, and MOS dimensions in a first approximation. Such MOS resistors enable high impedance at concurrently low stray capacitance in small signal domain. The complete amplifier chain inclusive LNA is optimized for low power consumption and achieves a gain-bandwidth-product of up to 4 GHz at just 1.8 µW.

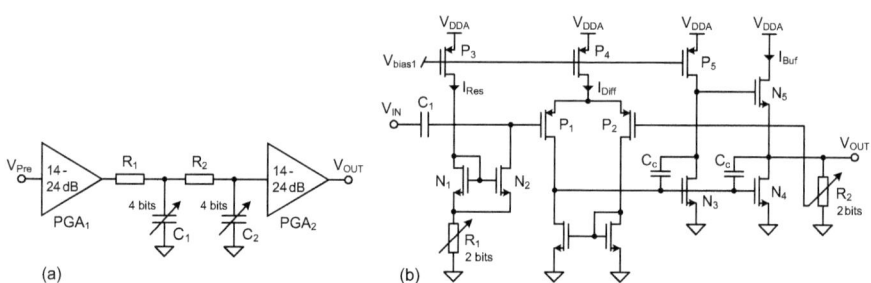

Figure 4.7: (a) Baseband amplifier chain with second order low-pass filter and (b), simplified schematic of baseband PGAs

Mixed-Signal Correlator

In order to enhance baseband SNR via reduction of noise bandwidth, the ultra-low power analog/mixed-signal correlation unit with switched-capacitor principle from figure 4.8(a) is deployed. It is designed for digital code sequences of 64 bit length and therefore includes 128 parallel stages because of oversampling for anti-aliasing of sampling circuit. The correlation scheme is operated by a single circular shift register that contains the digital code sequence of

1 bit resolution. Each register tap represents a bit-shifted version of the original code sequence. It is filtered first by a matched filter decoder that controls the corresponding correlator stage. The logical filter characteristic is illustrated in the timing diagrams of figure 4.8(b). The oversampled code sequence is divided into positive weights of $+1$ for positive code bits, negative weights of -1 for zero code bits, and zero weight at bit transitions. This way, control signals for the SC low-pass filter structure are generated according to sign and magnitude of the digital filter output. So the integration switches $SW_{P,0}$ and $SW_{N,0}$ are activated according to the actual weight of the oversampled code sequence. The analog baseband signal V_{BB} is sampled via capacitor $C_{S,0}$ during a first clock phase. Depending on polarity of the actual code bit, the sampled charge is then transferred in a second phase to either $C_{P,0}$ for positive bits, $C_{N,0}$ for negative bits, or even no charge is transferred for bit transitions. Since the capacitance ratio of $C_{P,0/N,0}/C_{S,0} = 700$, up to 700 bit can be averaged and that means more than 10 consecutive code sequences may be processed and accumulated within these two low pass filters. The profit over normal correlation length of code sequence is increased coding gain of up to 34 dB without raise of complexity or power consumption of circuitry. Currently, the integration capacitors are 1.0 pF, but they may also be larger for even higher coding gain. The correlation result is represented by the voltage difference between the integration capacitors $C_{P,0}$ and $C_{N,0}$, so signal evaluation and threshold comparison is achieved via comparators and a digital-to-analog converter simply by application of the decision-threshold potential at the negative connection of $C_{N,0}$. Much power can be saved, because the complete evaluation circuitry (comparators and digital-to-analog converter (DAC)) is power-gated and duty-cycled heavily. This is possible without aliasing effect because of the major reduced signal bandwidth at the integration capacitors $C_{P,0}$ and $C_{N,0}$. A wired-OR connection of the open drain comparator outputs of each of the 128 correlation stages and a latched wake-up interrupt status lowers power consumption further. The 7-bit DAC for threshold generation contains a binary weighted capacitive network with 2^7 equal capacitors and a voltage buffer amplifier with shut down option.

Power consumption of the complete correlation unit is only 490 nA at supply voltage of 1.0 V and an analog sample rate of 200 kS/s. The main reasons for it are low gate drive of the small analog switches, a small sample capacitor $C_{S,0} = 1.4$ fF, power-gated evaluation of the correlation result, and power-optimized full-custom design of the circular shift register and the matched filter decoder.

Figure 4.9(a) shows the positive edge triggered D-flip-flop for the circular shift register. It is based on two master/slave latches, has parallel load inputs, supports fully static operation and is designed with low clock-input load. When compared to equivalent D-flip-flops from digital standard library, this highly power-optimized cell consumes only 24 % for intended application.

Implementation

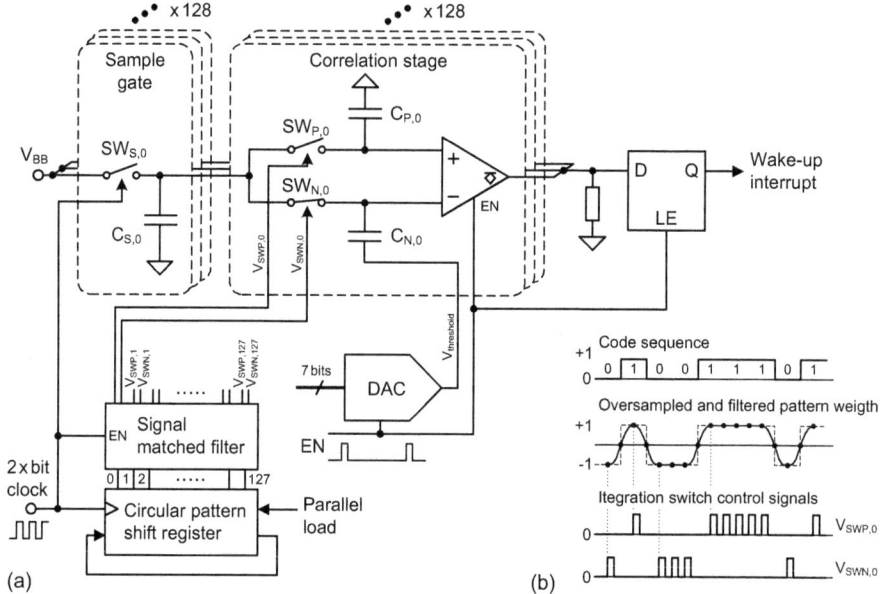

Figure 4.8: Block diagram of (a) proposed mixed-signal correlation unit and (b), illustration of matched filter output signals for adequate control of integration switches

Figure 4.9(b) depicts the schematic of two consecutive outputs of the matched filter decoder that controls the integration switches of the correlation stage. Extra small transistors for minimizing clock load are allowed because of the relaxed speed requirements. Thank to these digital full-custom cells, power consumption of the circular shift register and the matched filter is just 32 % of the total demand for analog correlation, otherwise it would predominate.

A/D Converters

To support advanced signal processing in digital domain, two different types of ADCs are implemented. The design of a converter that is based on PWM is illustrated in figure 4.10. It has inherently high amplitude resolution because of a direct magnitude-to-time conversion principle. So the pulse width of the digital output equals the sampled voltage magnitude and the final resolution is then defined off-chip via time-domain sampling of pulse width.

At the rising edge of the trigger input, the CMOS switch consisting of N_1 and P_1 is opened, the output signal OUT_{PWM} is set, and the input voltage IN_{Analog} is sampled within C_S first. Then, the previously discharged integration capacitor C_{INT} is charged linearly with constant current

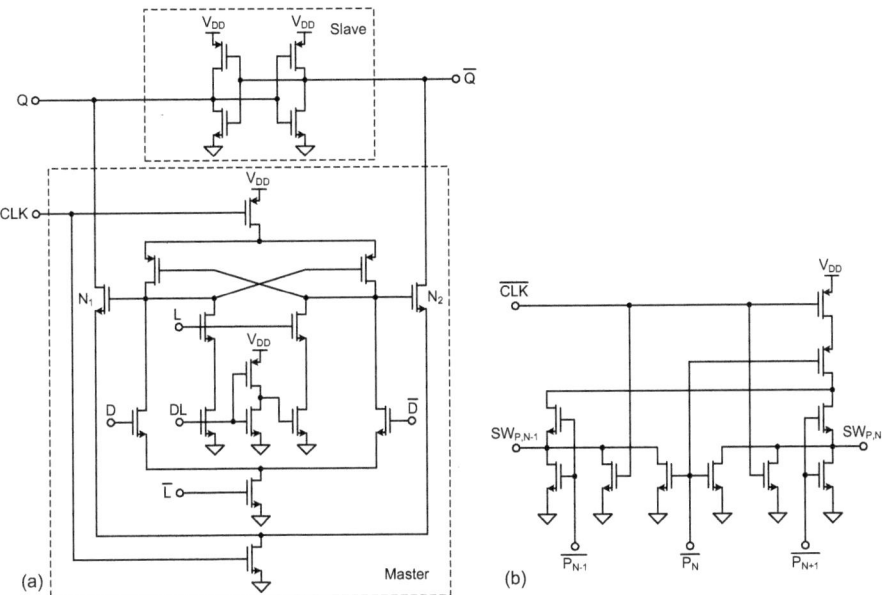

Figure 4.9: Digital low-power full-custom cell designs: (a) D-flip-flop cell and (b), matched filter decoder cell for digital correlation pattern bits P_N

I_{Charge} until the voltage comparator detects a match of sampled voltage V_{Hold} and V_{Ramp}, and resets the PWM output again. The maximum pulse with is equivalent to reference voltage and can be adjusted digitally via a tunable integration capacitor to support a wide range of sample rates.

Alternatively, the SAR ADC with resolution of 3 bit from figure 4.11 is implemented. Its digital data is clocked out serially on both edges of clock input. At the first positive edge, a conversion is triggered and the serial data output is always sampled high for synchronization of data stream. At the next negative edge, the MSB (bit 2) can be sampled at data output. Bit 1 and least significant bit (LSB) (bit 0) of the conversion result are read at consecutive rising and falling clock edges until the next conversion cycle is started with rising edge. As a consequence of this double data rate scheme, clock frequency has to be only twice the sample rate and thus, power is saved. The input sampling stage and the voltage comparator of the schematic are similar to those from the PWM ADC. Additionally, a digital state machine controls the capacitor ladder of the 3-bit DAC that generates V_{DAC} for comparison to V_{Hold}. According to considerations from [Con01], the design incorporates stray capacitance for high linearity.

Implementation

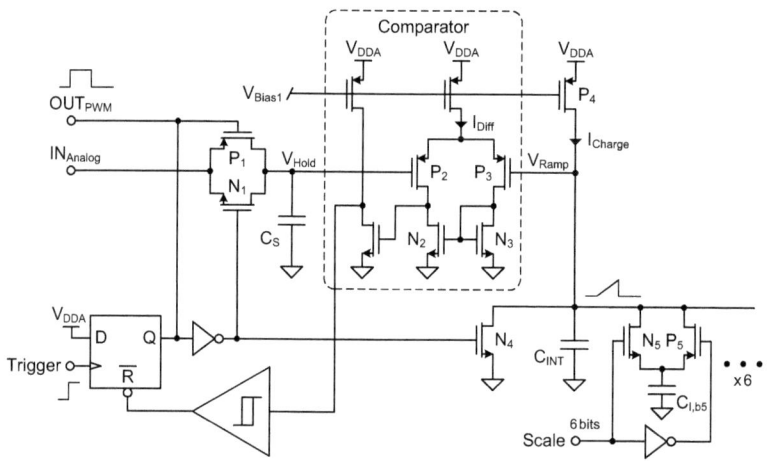

Figure 4.10: Simplified schematic of analog-to-digital converter with pulse width modulation (PWM) output

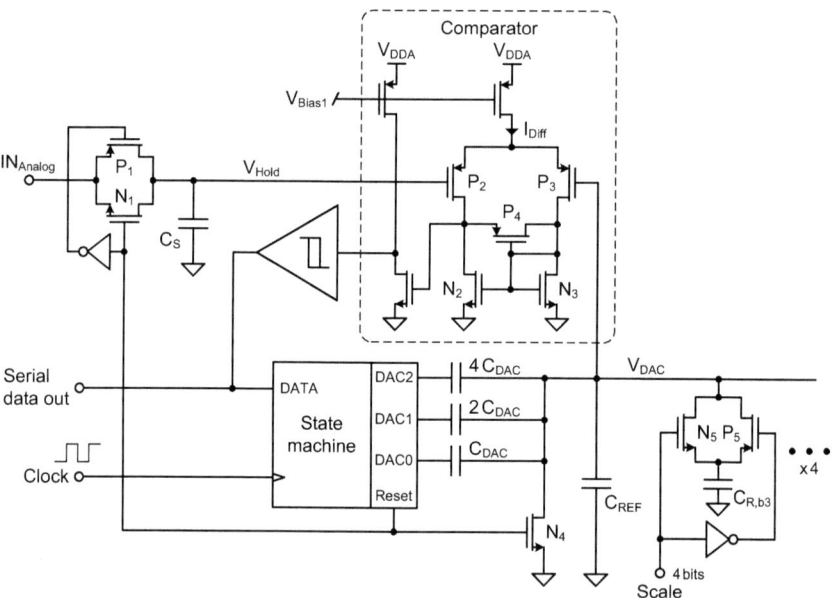

Figure 4.11: Simplified schematic of analog-to-digital converter with serial data output based on 3-bit successive approximation principle

Schmitt trigger

Schmitt triggers are very important components for low power interfacing from analog to digital domain. Slowly rising or falling signals from comparator outputs with direct connection to digital inputs would cause high risk for oscillation and enormous power consumption. So rise time has to be shortened by an additional gain stage with positive feedback. Common principles for Schmitt triggers utilize the substrate effect [ZSA03] and/or consume lots of current near their switching thresholds such as circuits from [KC01]. The schematic of the proposed design is shown in figure 4.12. It basically consists of three CMOS inverters and an output buffer. The essential criterion is positive feedback via P_4 and N_4. All MOS transistors are very small and can be scaled easily according to particular speed versus consumption requirements. Typically, the hysteresis thresholds are at $\frac{1}{3}$ and $\frac{2}{3}$ of supply voltage.

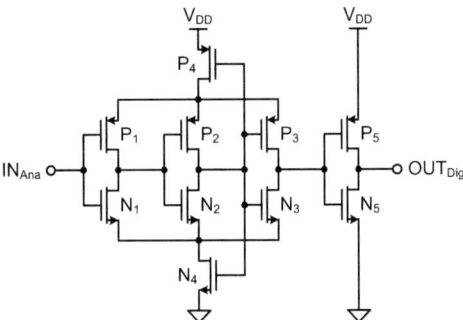

Figure 4.12: Schematic of proposed low power Schmitt trigger

Voltage and Current Reference

Nearly every analog circuit component needs bias current for operation. So a common current reference with multiple outputs is used to generate bias for the whole WuR ASIC in order to avoid matching problems in case of local bias creation. Figure 4.13 illustrates the implemented circuit for one 50 nA output and four 20 nA outputs. The core of the current reference is a self-biased beta-multiplier structure that contains of N_1, N_2, N_3, and P_1, P_2, P_3. All MOS transistors operate in weak inversion. Thanks to symmetry of transistors and voltage potentials, currents I_1 and I_2 are controlled to be almost equal over a large temperature range. So this structure has excellent power supply rejection ratio. Small cascode transistors N_9 and N_{11} enhance output impedance and force capacitive decoupling to PMOS devices. MOS capacitor P_4 stabilizes the feedback loop. Transistors $N_1 - N_9$ have equal gate length for adequate matching, but

Implementation

gate width is different and noted at the respective device. So voltage V_{PTAT} is derived from equation 3.4, results in

$$V_{PTAT} = V_T \ln \frac{W_{N2}}{W_{N1}} = V_T \ln 8 = 53.4\,\text{mV} \qquad (4.2)$$

at 25 °C, and is proportional to absolute temperature. Finally, the poly-silicon resistor R defines the reference current and hence, it is tunable over the full range of fabrication variability for compensation purpose. In order to ensure a small resistor R for high bandwidth and low area occupation, the largest reference currents I_{50n} and $I_{20n,1}$ are added to $I_{Res} = I_2 + I_{50n} + I_{20n,1} = 90\,\text{nA}$ at 25 °C. Save startup of self-biased circuits under all possible conditions is challenging for low power consumption. The proposed startup circuit from figure 4.13 includes a MOS capacitor P_5 that turns on N_{12} and N_{13} when supply voltage V_{DD} rises and thus, startup current is injected into the beta-multiplier. In steady state, N_{11} charges P_5, and N_{13} is turned off again such as proposed similarly in [KWM03]. An additional trick to reduce leakage current of N_{13} is to lift its source potential via P_6. Therefore accuracy is enhanced further, especially at elevated temperature.

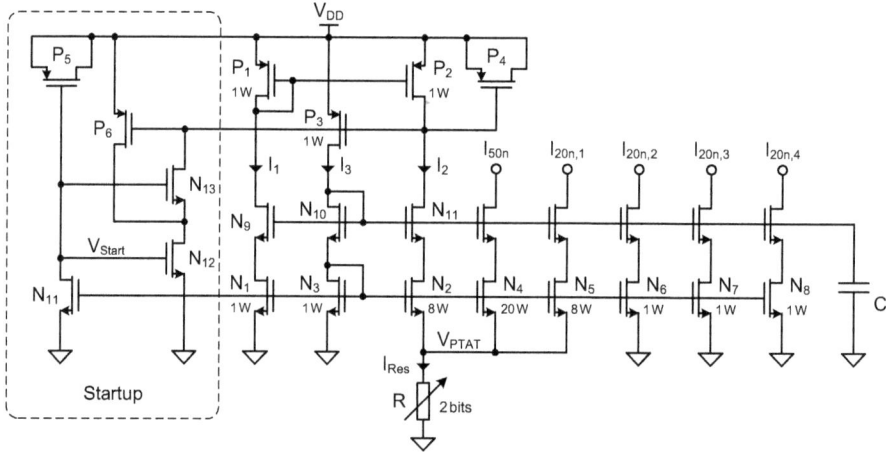

Figure 4.13: Schematic of low-power current reference with 5 outputs: The design implements a self-biased beta-multiplier structure with weak-inversion MOS transistors and includes the startup circuit.

The schematic of the novel ultra-low power voltage reference from figure 4.14 has a similar basic structure compared to the previously described current reference. Beta-multiplier and startup circuit are almost equal. The additional chain of 6 series connected MOS voltage dividers

Implementation

compensate the negative temperature coefficient of the diode connected NMOS transistor N_{11a} to achieve stable reference voltage V_{REF}. Each of the bias currents $I_1..I_9$ and also the two reference current outputs are targeted to be 2.0 nA with PTAT characteristic. So the overall consumption including startup circuit is then 22.2 nA in steady state. The trick to get a feasibly low resistor R in spite of nanoampere is that bias currents of all MOS dividers as well as reference currents are collected to I_{RES}, and also the resistor voltage is reduced to $V_{PTAT0} = V_T \ln \frac{W_{N2}}{W_{N1}} = 28.2$ mV at $25\,°C$. Then $R = \frac{V_{PTAT0}}{I_{RES}} = 1.53$ MΩ. According to equation 3.3, the full PTAT voltage results in $V_{PT6} = V_{PTAT0} + V_T \left(\ln \left(1 + \frac{1}{1}(1 + \frac{5}{1}) \right) + \ln 6 + \ln 7 + \ln 7 + \ln 6 \right) = V_T \ln 259308 = 320.2$ mV. A load capacitor $C = 1.0$ pF suppresses high frequency noise and the remaining noise voltage $V_{N,RMS} = 1.1$ mV is derived from simulation. Other advantages of this design are the support of very low supply voltage nearly down to V_{REF} and concurrently fast startup.

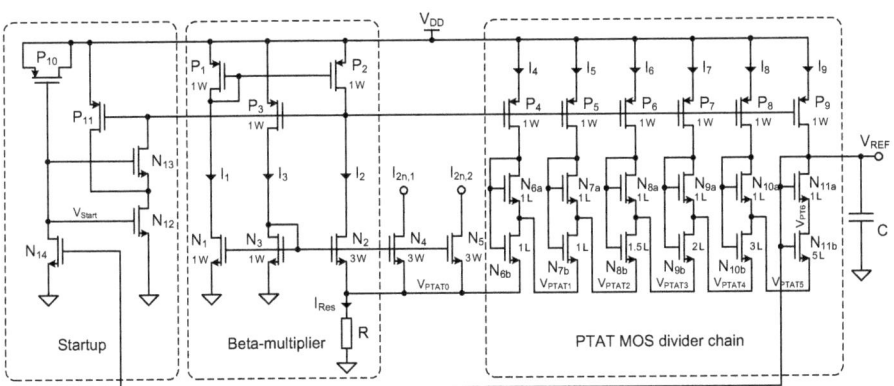

Figure 4.14: Simplified schematic of novel ultra-low power voltage reference
PTAT: proportional to absolute temperature

4.2.3 Layout

Chip layout can have major impact to circuit properties. In order to reach high performance and simultaneously low power consumption, the ASIC layout is done manually in a large extent. All analog and RF parts as well as relevant digital parts with predominant power consumption are fully designed by hand down to device level. Libraries are used only for ESD structures, I/O blocks and I/O pads. Cells of the digital standard library are applied just for digital circuits with negligible contribution to total power demand.

Implementation

The main issues for a good layout are adequate analog transistor matching for low offset voltages, consideration of wiring resistance, and – especially for low power designs – minimization of stray capacitance and coupling. For certain critical circuit parts, a post-layout simulation can include resistance and also coupling capacitance of wiring. With the help of layout-extracted back-annotation to simulation netlist, power consumption of digital designs or performance of analog circuits are checked against previous assumptions. Additional test pads on top of the chip ease debugging via a wafer prober if necessary.

Figure 4.15 shows the layout of the complete WuR ASIC. Its total size is $1.15 \times 1.0\,\text{mm}^2$. Chip area is dominated by space occupation of the surrounding pad-ring and can be reduced significantly when it is integrated into a system-on-chip. The pad-ring includes structures for ESD protection as well as bond pads of $50 \times 50\,\text{µm}^2$ for wire connection. The left half of the ASIC contains the analog frontend with baseband amplifier chain, filter, and current reference. It is separated from the digital domain VDDD/VSSD by means of an additional analog substrate connection VSSA/VSSRF to decouple noise. The dominant dark area represents MOS capacitors for decoupling, and the periodic structure in the right half depicts 128 stages of the mixed-signal correlation unit. SPI, configuration registers and test circuits require comparatively little space at the bottom chip edge. When extending a main receiver by addition of the proposed wake-up receiver design, actual semiconductor demand would increase only about $0.26\,\text{mm}^2$ without extra pads.

Figure 4.16 shows the layout of the ultra-low power voltage reference up to layer *Metal2* and without fill structures. The dominant layer is polysilicon and either gate connection, or it represents a high impedance resistor in meander shape in the left half of the drawing. Two arrays of PMOS and NMOS transistors include dummy devices that surround the actively used transistors for appropriate device matching, so variation of MOS threshold voltage is minimized when ambient neighborhood is almost equally. The layout size of the voltage reference is just $107 \times 46\,\text{µm}^2$, whereupon around half of the area is occupied by the $1.53\,\text{M}\Omega$ resistor R for tolerance reduction due to geometrical process variations.

The layout in figure 4.17 depicts main parts of the anlog/mixed-signal correlation unit. Each correlation stage is implemented with high symmetry for low offset voltage of output comparators and high grade of cancelation for layout parasitics. One can see from the picture that the integration capacitors occupy most space of the total correlation unit. Layout of switches and the shift register with matched filter decoder are optimized for minimum stray capacitance in order to achieve low power consumption of the digital parts. The regular structure within the layout of the 7-bit DAC contains 128 identical capacitors. They are constructed of a shielded layer stack of polysilicon and layers *Metal*1..4 for linear charge characteristic. Comparatively low area is required by the digital configuration registers and the SPI interface.

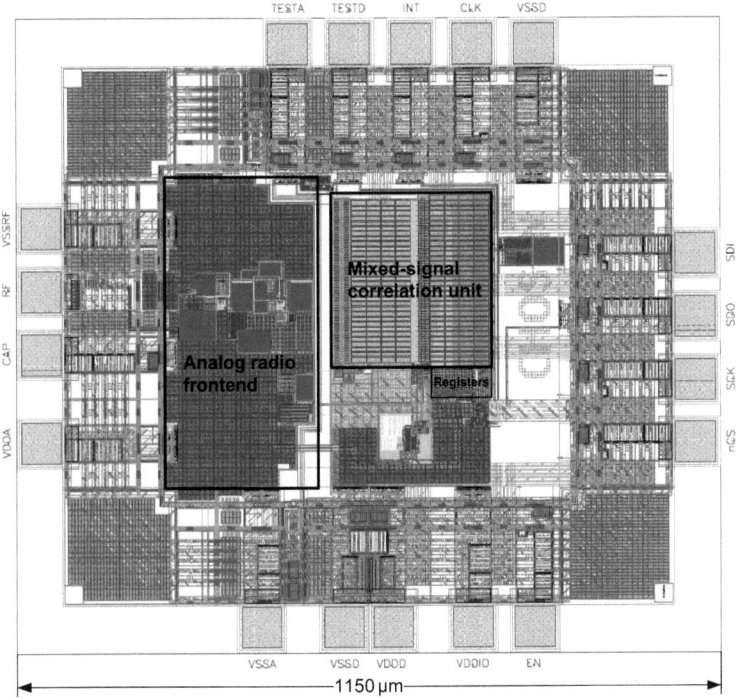

Figure 4.15: Layout of complete WuR ASIC

Without pad-ring and its ESD structures, the WuR's layout contains more than 11600 MOS devices for analog purpose as well as around 1700 transistors from pre-layouted standard library used in digital domain. Thereof, the mixed-signal correlator itself consists of approximately 6500 MOS devices. All analog and main parts of the digital circuitry are designed manually. Also layout is optimized for low static and low dynamic power consumption. This comprises SPI interface, configuration register cells with read and write access, and I/O drivers.

4.2.4 Register Map

The WuR design contains 15 registers for configuration and control purpose. All of them are readable and writable, and reset to their default value at power-up. Table 4.2 lists the full register map with register name, register address, and initial bit values.

Registers $PAT0..7$ define the digital 64 bit code pattern for the mixed-signal correlation unit.

Implementation

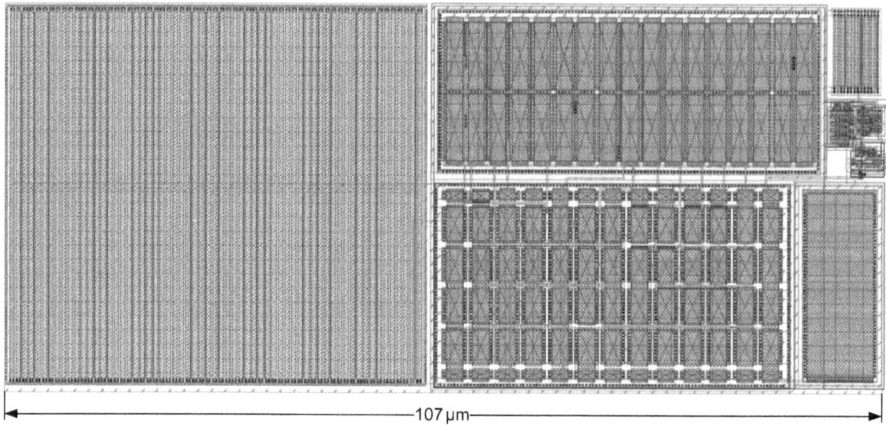

Figure 4.16: Layout of proposed ultra-low power voltage reference: The main layers shown are diffusion, polysilicon, $Metal1$, and $Metal2$.

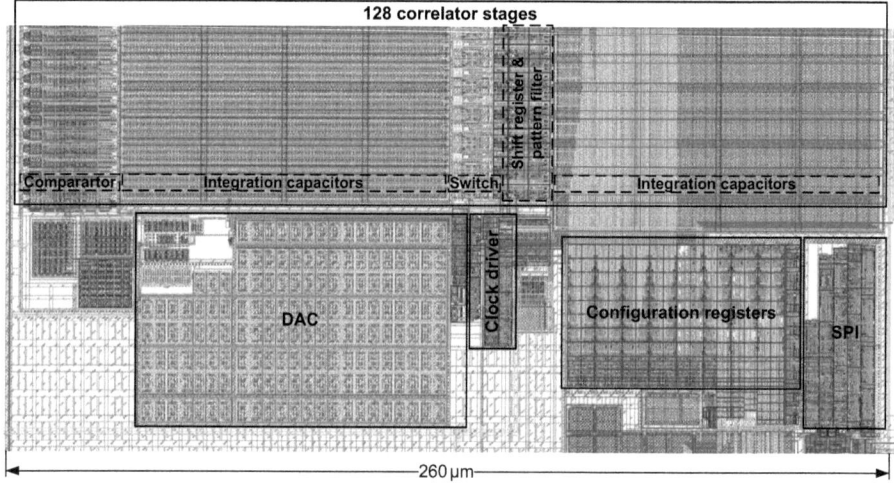

Figure 4.17: Layout of mixed-signal correlation unit

The bit sequence is fully user programmable and it is utilized for circular correlation with baseband signal, whereas MSB (PT63) is the first sequence bit. The $TEST$ register is used for testing, possibly debugging and characterization purposes only. It can enable a digital and also an analog output buffer for one of several multiplexed input channels. $CTIM$ bit of correlation register $CORR$ selects correlation length. If set, the interval for correlation is equivalent to

Register	Addr.	Bit 7	Bit 6	Bit 5	Bit 4	Bit 3	Bit 2	Bit 1	Bit 0
PAT0	0x00	PT7	PT6	PT5	PT4	PT3	PT2	PT1	PT0
	default	1	1	0	1	0	1	1	0
PAT1	0x01	PT15	PT14	PT13	PT12	PT11	PT10	PT9	PT8
	default	1	0	1	1	0	1	0	0
PAT2	0x02	PT23	PT22	PT21	PT20	PT19	PT18	PT17	PT16
	default	0	0	1	0	0	1	1	1
PAT3	0x03	PT31	PT30	PT29	PT28	PT27	PT26	PT25	PT24
	default	0	0	1	0	0	1	0	0
PAT4	0x04	PT39	PT38	PT37	PT36	PT35	PT34	PT33	PT32
	default	0	1	0	0	0	0	0	1
PAT5	0x05	PT47	PT46	PT45	PT44	PT43	PT42	PT41	PT40
	default	1	1	0	1	1	1	1	0
PAT6	0x06	PT55	PT54	PT53	PT52	PT51	PT50	PT49	PT48
	default	1	1	1	1	0	1	1	1
PAT7	0x07	PT63	PT62	PT61	PT60	PT59	PT58	PT57	PT56
	default	0	0	0	1	0	1	1	0
TEST	0x09	TDEN	–	–	VREN	TAEN	–	ACH1	ACH0
	default	0	–	–	0	0	–	0	0
CORR	0x0A	CTIM	THRE6	THRE5	THRE4	THRE3	THRE2	THRES1	THRE0
	default	1	0	0	1	1	0	0	0
ADCFG	0x0B	ADSEL	–	ACFG5	ACFG4	ACFG3	ACFG2	ACFG1	ACFG0
	default	1	–	1	0	1	1	0	0
OPTION	0x0C	GEN	–	COREN	ADEN	–	–	IOD	IAH
	default	0	–	1	0	–	–	0	0
AMP1	0x0D	AMOD	–	GLF1	GLF0	GB1	GB0	GA1	GA0
	default	1	–	1	0	1	1	1	1
AMP2	0x0E	BIAS1	BIAS0	–	BWLL	BWLH3	BWLH2	BWLH1	BWLH0
	default	1	0	–	0	0	1	1	1
RFCFG	0x0F	RFEN	RFLD	–	RFCT4	RFCT3	RFCT2	RFCT1	RFCT0
	default	1	0	–	1	1	1	1	1

Table 4.2: Register map of WuR ASIC implementation

a duration of approximately 700 bit, otherwise interval equals approximately 300 bit of transmitted sequence, whereupon longer correlation period results in increased receive sensitivity. $THRE6:0$ bits define the configurable decision-threshold for interrupt event generation. The LSB equals a nominal voltage step of 4.9 mV. It is programmed according to optimal balance between acceptable false wake-up rate that is caused by noise, and receive sensitivity. If set, the $ADSEL$ bit of $ADCFG$ register selects the 3-bit ADC with successive approximation principle, otherwise the converter with PWM output is chosen. For both types of ADCs, bits $ADCFG5:0$ determine power mode as well as voltage reference for A/D conversion. The global enable bit GEN of the $OPTION$ register enables the complete analog frontend and also accordant baseband signal processing unit to put the WuR in active mode. When this

Implementation

bit is cleared, the WuR goes to power-save mode and only SPI interface and configuration registers are powered for content retention. $COREN$ bit enables the mixed-signal correlation unit if set, and $ADEN$ bit enables the selected type of ADC. If IOD is set, the driver of the interrupt output is configured for open drain operation, otherwise it has a push-pull output stage. If IAH bit is set, the INT output pin is active high, or active low contrariwise. The $AMP1$ register contains the $AMOD$ flag for calibration and adjustment of reference voltage for optimal A/D conversion, and the $GLF1:0$ bits for configuration of the LNA's low-frequency cutoff between $1.0 - 3.7\,\text{kHz}$. This offers the possibility to suppress flicker noise of the first gain stage. Bits $GB1:0$ and $GA1:0$ select gain of the PGAs with options for $14/19/24\,\text{dB}$. The nominal reference current of $50\,\text{nA}$ is tunable by $+20/-25\,\%$ via $BIAS1:0$ for compensation of worst case technology variability. The $BWLL$ bit limits lower bandwidth of the PGAs for shortened settling time and $BWLH3:0$ bits determine upper cutoff frequency of the anti-alias filter in a range from $25\,\text{kHz}$ to $560\,\text{kHz}$. In $RFCFG$ register, a cleared $RFEN$ bit shorts the RF input and thus, most of RF power is reflected. So the remaining baseband noise can be used for calibration of amplifier gain. When $RFLD$ bit is cleared during normal operation, an automatically controlled RF magnitude limiter prevents the envelope detector and the LNA from strong overdriving at high receive signal strength. With the help of bits $RFCT4:0$, the equivalent input capacitance of the RF pin can be tuned by some $\pm 5\,\%$ for compensation matching network tolerance. A binary value of 0b11111 results in minimum capacitance, while 0b00000 gives maximum capacitance and thus minimum resonance frequency in conjunction with inductor L_1 from figure 4.5.

4.3 System Simulation

Most of the simulation tasks for ASIC design is done with the help of the Cadence tool set [36] in the course of circuit design. This comprises mainly circuit simulation with very accurate models of CMOS devices, but also behavioral simulation on system level. Such abstract models are implemented in SpectreHDL that supports mixed-signal modeling at flexible granularity and ensures short execution time at appropriate accuracy. The second high level tool that is used for evaluation of the design for baseband signal processing is MATLAB [37]. Common circuit simulation tasks include transient analysis with optional steady state evaluation, DC-analysis, AC-analysis, and noise analysis. The impact of temperature characteristic is investigated, and also analysis at the technology's corner conditions is appraised via parameter stepping to ensure high reliability and minimize probability for failure because of variation during the fabrication process.

This section presents final simulation results of primary WuR system performance.

4.3.1 Analog Frontend

Figure 4.18 shows the simulated frequency response of the RF matching network. It contains magnitude characteristic of voltage gain $|S_{21}|$ and input return loss $|S_{11}|$. The simulation model already includes parasitic capacitance of inductor windings, loss of off-chip inductors, as well as input impedance of ESD structures and the proposed MOS detector circuit with layout-extracted on-chip loss. Thanks to the design of a high-quality resonance circuit, very high voltage gain of 27 dB and low bandwidth of 18 MHz is achieved. Consequently, the benefit is high sensitivity and increased suppression of interferers at adjacent frequencies. Hence, the drawback of this good frequency selectivity is the need for appropriate adjustment of center frequency to 868 MHz. Then, less than 1 % of RF input power is reflected.

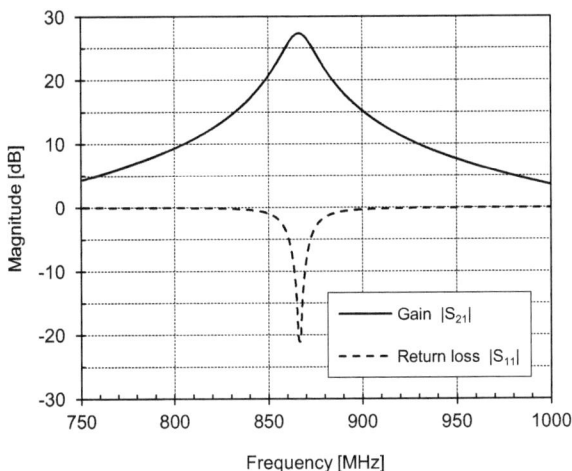

Figure 4.18: Simulated frequency response of matching network: S-parameter magnitude of voltage gain and RF input impedance

The simulated frequency response of the entire baseband amplifier chain is illustrated in figure 4.19. It depicts gain of the low noise preamplifier with different options for LF cutoff via the belonging dashed lines, as well as additional gain of both PGAs also with gain options of 14/19/24 dB for each amplifier. The elevated gain around some 200 kHz results from the second gain stage of the preamplifier. This way, the advantage of low spectral noise density (see also figure 4.20) in the concerned frequency range is emphasized. A maximum gain of up to 80 dB, bandwidth from 100 Hz up to 450 kHz, and flexible configuration options provide sufficient room for effective baseband signal conditioning.

Figure 4.20 shows the input-referred equivalent spectral noise density of the entire amplifier

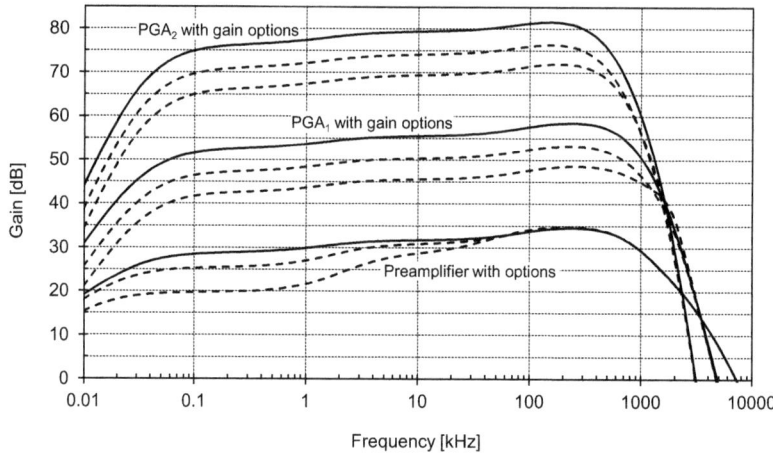

Figure 4.19: Frequency response of amplifier chain with various options for gain adjustment
PGA: programmable gain amplifier

chain for maximum gain configuration. At frequencies below 10 kHz, $1/f$-characteristic of flicker noise is visible. Above, thermal noise with constant spectral density covers the frequency range with best performance, until noise density increases again because of limited the gain bandwidth product. With the help of equations 3.17 and 4.1, and the given detector bias current $I_{DET} = 1\,\mu\text{A}$, the lower boundary of thermal generated spectral noise density becomes

$$V_N = 2kT\sqrt{\frac{n}{qI_{DET}}} = -33.0\,\text{dB}\mu\text{V}/\sqrt{\text{Hz}}. \qquad (4.3)$$

This lower physical boundary is exceeded only about 1 dB in implementation mainly because of additional noise of subsequent gain stages. Consequently, the transmitter's bit rate is chosen to operate the WuR's amplifier chain predominantly in this beneficial frequency range with low noise figure.

4.3.2 Mixed-Signal Correlation

Functional behavior and mathematical performance of the proposed analog/mixed-signal correlation unit is evaluated via a high level MATLAB model, while power consumption is estimated using circuit simulation. This approach supports detailed analysis of the intended algorithm. The concept is proven, parameters are optimized, and a reasonable range for configuration options is identified by the functional model at first.

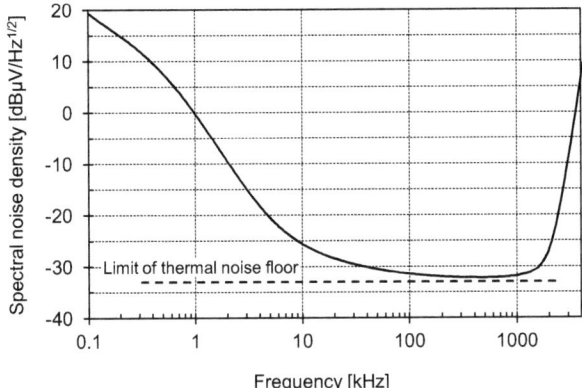

Figure 4.20: Input referred equivalent spectral noise density of baseband amplifier chain

Figure 4.21(a) shows a block diagram of a signal processing chain that emulates behavior of the proposed correlation scheme and analyzes its performance. The first block generates a well defined baseband signal with given input signal-to-noise ratio SNR_I and signal power P that behaves almost equally when compared to the analog baseband signal from analog frontend. Therefore, the digital 63 bit code pattern and also white Gaussian noise are generated, filtered with bandpass characteristic to emulate the baseband amplifier's frequency response, weighted according to SNR_I, and added. An amplitude limiter models overdrive effects of infrequently noise peaks. The second block implements the actual correlation algorithm. It contains the matched filter decoder with multiplier weights of $+1, 0, -1$, and accumulates the positive output as well as the negative result such like the proposed implementation with SC low pass filters from figure 4.8. Finally, the output signal-to-noise ratio SNR_O is calculated in a two-step approach. First, the nominal baseband signal consisting of desired signal and noise is fed through the correlator, and in a second step, just the exactly same noise is processed. This way, the SNR at the output can be derived as $SNR_O = \frac{P_O(S+N)}{P_O(N)} - 1$.

Figure 4.21(b) illustrates baseband signal at the correlator input for a typical $SNR_I = -20\,\text{dB}$. The accordant digital pattern component within this signal is no longer recognizable. In contrast, after correlation $SNR_O = +14.8\,\text{dB}$, and the amplitude step resulting from activation of desired signal can be evaluated easily in figure 4.21(c) despite the residual noise. The impact of different code sequences to correlation performance is marginal.

The normalized time constant of the SC low pass filters is 700, so up to 700 samples of baseband signal are accumulated and averaged. The resulting benefit from reduction of noise magnitude is then $\sqrt{700}$ and thus 28.5 dB, if statistical independency is assumed for a first approximation.

Implementation

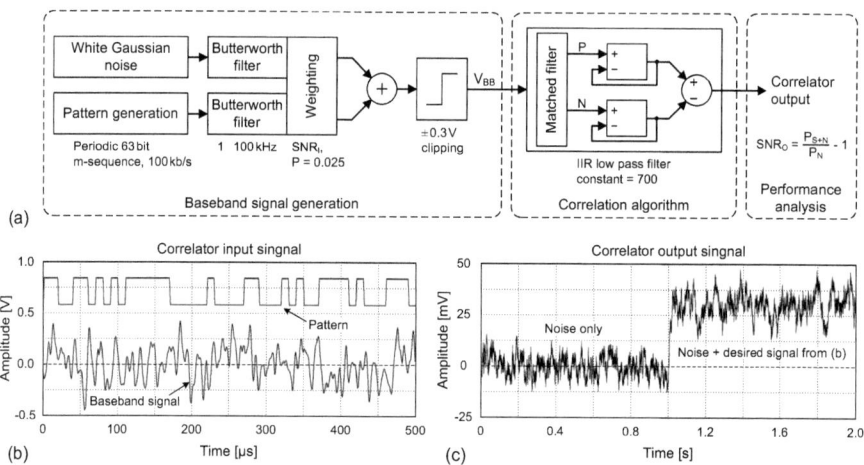

Figure 4.21: Mixed-signal correlation: (a) block diagram of simulation model, (b) emulated baseband signal with $SNR_I = -20\,\text{dB}$ and (c), corresponding correlation result without and with desired input signal and different time scale
SNR: signal-to-noise ratio

Furthermore, the desired signal is gained about a factor of 2 in amplitude or equivalently 6 dB, because the correlation result equates peak-to-peak amplitude of the desired input signal. Then theoretic coding gain is in total 34.5 dB. However, detailed simulation of the correlation stage leads to SNR gain of 34.8 dB for a noise bandwidth of 100 kHz and a sample rate of 200 kS/s. When noise bandwidth is reduced below 100 kHz, coding gain suffers. But one can argue that for that case, sample rate is chosen too high and unnecessary power is wasted. The optimum adjustment for maximum coding gain is a sample rate twice the noise bandwidth, and this still guarantees anti-aliasing. For a bit rate of $R = 100\,\text{kbit/s}$, 128 parallel correlation chains that can trigger an interrupt, and for an allowed false wake-up rate of $FWR = 10^{-3}/\text{s}$, the bit error ratio has to be lower than $BER \approx \frac{FWR}{R} = 10^{-8}$. According to figure 3.12, the decision threshold for interrupt generation then must be at least 12 dB above noise level.

4.3.3 Key Parameter Summary

The implemented ASIC design of an ultra-low power wake-up receiver solution basically combines a simplified radio frontend with a novel analog correlation technique in baseband domain in order to compensate the frontend's sensitivity penalty via high coding gain and increase the SNR. So the aim is to exploit preferably all possibilities to maximize receive sensitivity at simultaneously minimum power consumption. The intention of an ASIC implementation close to the physical limitations of the concept is realized successfully.

Main characteristics, expected results and essential design parameters from simulation are listed in table 4.3 for comparison.

Parameter	Expected result	Conditions
Wake-up sensitivity	< -70 dBm	10 ms correlation time, 100 kbit/s raw bit rate
Latency	< 10 ms	Reduced latency at high signal strength
Carrier frequency	868 MHz	Off-chip defined
False wake-up rate	< 10^{-3}/s	Decision threshold for 12 dB SNR
Power consumption	< 3 µW	Total wake-up receiver
	≈ 2 µW	Analog frontend
	< 1 µW	Mixed-signal correlation unit
Supply voltage	1.0 V	Analog/digital core
	1.2 V – 3.6 V	I/O interface
Operating temperature	-40 °C – 125 °C	Automotive range
Chip area	< 300 × 300 µm^2	Active area without pads

Table 4.3: Expected characteristics of the proposed WuR concept from simulation results

Design decisions and circuit diagrams of the main WuR components have been presented and discussed in this chapter. Not everything of the proposed functionality has been realized on chip due to lack of resources. Advanced power management with PLL, switching regulators, and on-chip digital signal processing are omitted for the first silicon. Nevertheless, the proposed WuR concept with auxiliaries for a full-fledged companion-chip solution with microcontroller support has been implemented – all with focus on ultra-low power consumption.

5 Measurement Results and Discussion

Careful evaluation of the implemented design and interpretation of deviations from expected or simulated results yield to valuable feedback for future improvements of design, or it can lead to refined simulation models with increased accuracy. So this chapter presents accordant measurement results with the implemented wake-up receiver (WuR) ASIC. Measured performance parameters of the whole WuR system and also of main subsystems are given, analyzed, and discussed. The results are compared with simulation and related work. Finally, a detailed summary of the achieved performance characteristics affords a precise overview about quality of this work and supports benchmarking of the implemented WuR concept.

5.1 ASIC Characteristics

5.1.1 Measurement and Test Environment

Figure 5.1 shows a photograph of the main development platform that is used for test setup and measurements. The WuR ASIC prototypes are assembled directly onto a PCB board (see also figure 6.1). Beside actual WuR functionality, it contains additional drivers and connectors to ease measurement setup. This WuR PCB is designed as plug-in module for the platform carrier board to allow flexible interconnection and wireless node emulation. Further plug-in modules are a FPGA board with soft-core compatible Spartan-3 device from Xilinx, a transmitter module with TDA5150 [4], a receiver module with TDA5240 [4], and a microcontroller board with XC886 from Infineon [4]. A USB to UART/SPI converter interface, LEDs, push-buttons and switches, and multiple clock options support application development and debugging. External devices, modules, or measurement equipment may be connected directly to FPGA I/Os via expansion connectors. This platform offers the possibility for full functional sensor node implementation.

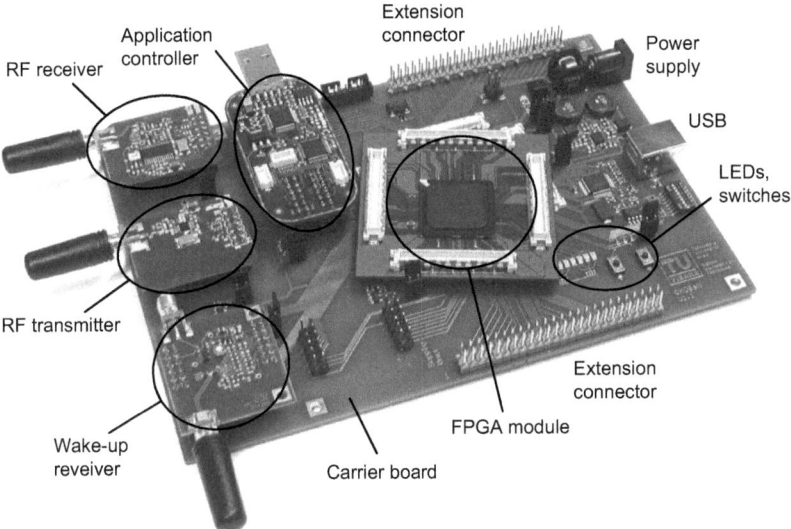

Figure 5.1: FPGA development platform from Vienna University of Technology for the CHOSeN research project [1]: The generic platform allows flexible and comfortable configuration and test of the WuR PCB module. The plug-in modules for the XC886 microcontroller, the main receiver TDA5240 and the transmitter TDA5150 from Infineon [4] can be connected via the Xilinx FPGA XC3S1000 with various options for extensions.

While analog frontend, mixed-signal baseband processing unit, and analog-to-digital conversion is implemented on the WuR chip, digital signal processing can be done within the FPGA. This allows flexible design and optimization with real environmental measurement data that include actual noise characteristic and behavior of radio channel. So development and measurement under realtime conditions is more efficient when compared to purely simulation.

Figure 5.2 depicts screenshots of the prepared software tools for comfortable control of the WuR module and transmitter module in order to accelerate device characterization. The WuR configuration tool provides full read and write access to the ASIC registers and is able to control all signals of the digital interface. The transmitter tool offers easy entry of transmit pattern and supports periodic transmission at configurable bit rate. Both software tools are connected to their corresponding hardware modules via a USB to SPI converter. This portable setup alleviates extensive performance tests with multiple nodes also in outdoor environment.

Measurement of temperature characteristics is done with the help of setup from figure 5.3. A thermoelectric module (TEM) is used for cooling and heating the PCB-mounted WuR dies

Measurement Results and Discussion

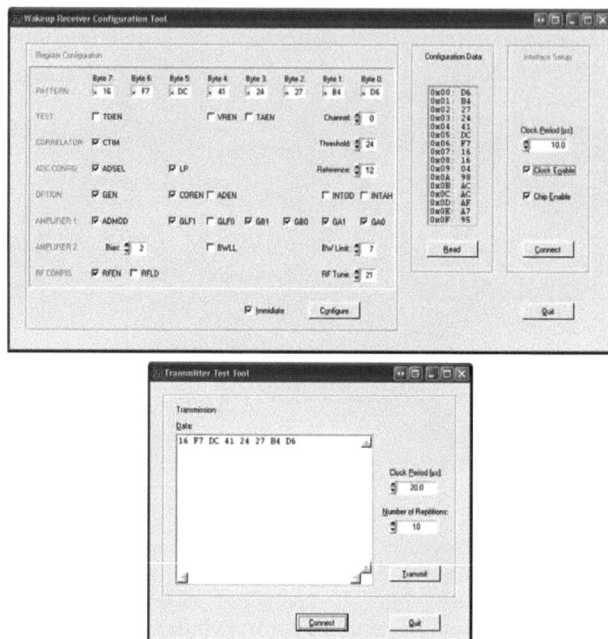

Figure 5.2: Screenshot of software tools for control of wake-up receiver module and transmitter module

down to -30 °C and up to +100 °C. The control loop for temperature regulation is operated by a MATLAB PC that steers the GPIB connected power source and evaluates actual device temperature with an on-board semiconductor sensor via the digital multimeter. Tight thermal coupling of ASIC, temperature sensor, TEM, and low thermal mass result in a short thermal time constant of around 6 s. With optimized control loop parameters, temperature profiles are passed through quickly and with high accuracy. Cooling down below ambient temperature leads to condensation of humidity and thus, ice crystals grow on the device under test when temperature falls below 0 °C. Since ice is a good electrical insulator, current measurement in nanoampere range is only critical between 0 °C and room temperature, where condensing water is fluid.

5.1.2 Analog Frontend

The main parts of the analog frontend are RF matching network, envelope detector, preamplifier, PGAs, and ultra-low power voltage reference. Most important measurement results of

Measurement Results and Discussion

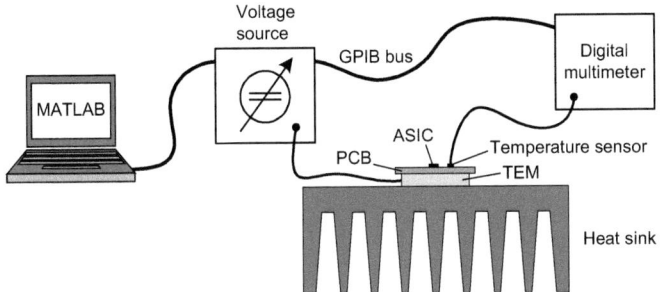

Figure 5.3: Setup for measurement of temperature characteristics: A thermoelectric module (TEM) is used for cooling and heating the PCB-mounted ASIC down to -30 °C and up to +100 °C.

their implemented designs are presented and discussed in the following charts.

Input Matching Network

One major impact to receive sensitivity is RF voltage gain as consequence of impedance transformation via the matching network. Figure 5.4 compares frequency response of measured voltage gain magnitude $|S_{21}|$ with simulation. At center frequency, the measured curve (solid line) is 6 dB below simulation result and also its bandpass filter characteristic is broadened. This deviation from simulation is the effect of increased loss within the resonance circuit. Causal research lead to the finding that equivalent series resistance of the MOS transistor's input impedance is not modeled in the used semiconductor technology. Hence, increased loss and reduced sensitivity comes from model inaccuracy. Nevertheless, impedance transformation still gains $|S_{21}|$ by 21 dB. When the off-chip SAW filter with a bandwidth of 11 MHz is included too, additional 2.2 dB of insertion loss has to be considered for the total filter characteristic, but out-of-band interference on adjacent channels such as GSM is cut off excellently. Margin for further improvement is low, because implementation is already optimized for a low-loss RF input path.

The Smith chart of figure 5.5 illustrates measured RF input impedance of the WuR between 600 MHz and 1.1 GHz. The short line represents the impedance characteristic of the ASIC's RF input and shows nearly ideal capacitive behavior in a frequency range from 600 MHz to 1.1 GHz. For a configuration of minimal input capacitance, impedance is 8.45 − j318 Ω at 868 MHz. The real part mainly comes from non-modeled input loss of the MOS transistor for envelope detection. When the matching network is included in measurement, the frequency response is matched to 50 Ω quite optimally at 868 MHz. Already moderate deviation from

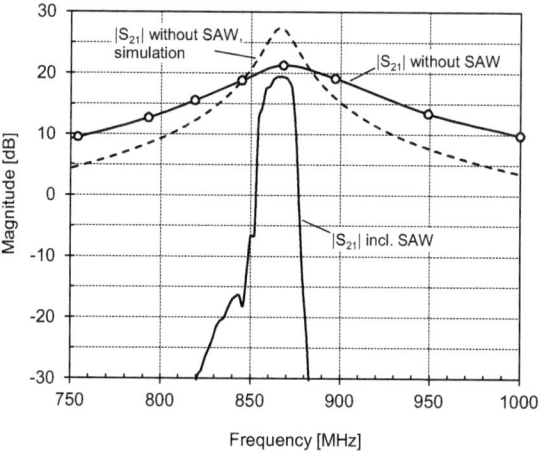

Figure 5.4: Voltage gain of matching network with and without off-chip surface acoustic wave (SAW) filter

this nominal center frequency leads to high reflection coefficients and confirms with the 3 dB-bandwidth of 55 MHz from figure 5.4.

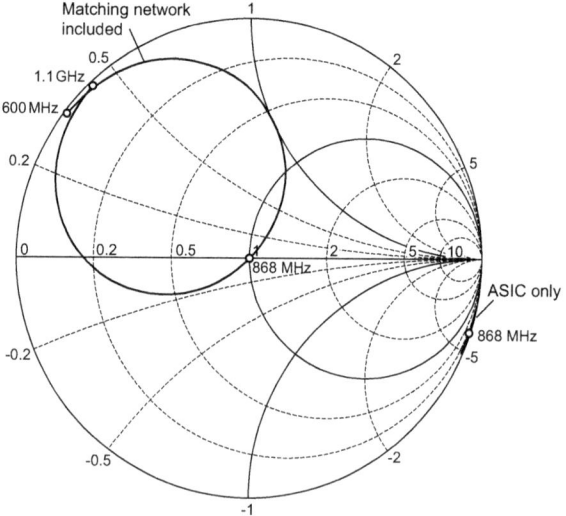

Figure 5.5: Smith chart: RF input impedance of WuR ASIC with and without off-chip matching network

Figure 5.6 illustrates an impedance-equivalent schematic of the WuR's RF input structure. Its values are calculated from impedance measurements and contain also parasitics from wire bonding, pads and ESD protection circuits. In receive mode, input characteristic is nominally 576 fF in series with about 8.45 Ω of loss. The two off-chip inductors match this to a 50 Ω antenna load at 868 MHz. For compensation of inductance variation, the RF input capacitance is tunable by ±5 % to adjust the resonant frequency for maximum sensitivity. If closed, the built-in RF switch can be used for noise calibration purpose in backend signal processing. Then, the ASIC's impedance is 118 – j45 Ω and it provides only 0.32 dB of return loss in a frequency range from 600 MHz to 1.1 GHz when the matching network is included. So nearly all RF power is reflected and this can be utilized to support antenna switching, if a shared antenna is used for the WuR and the main transceiver.

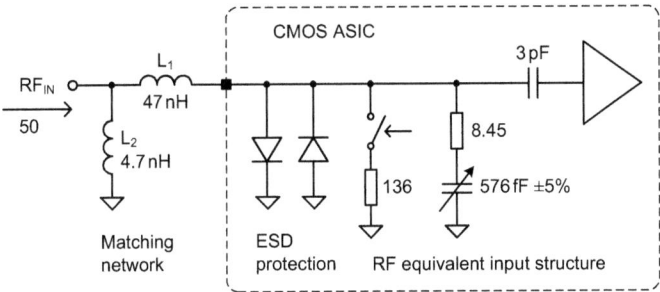

Figure 5.6: WuR's equivalent RF input structure based on impedance measurements at 868 MHz

Figure 5.7(a) depicts measured input return loss and calibration range for center frequency. Since $|S_{11}| < -35$ dB at all three resonance frequencies for minimum, nominal, and maximum tuning options, impedance is matched very well and can be tuned about 42 MHz. Part (b) of figure 5.7 shows linear and strictly monotonic characteristic of the implemented frequency calibration possibility.

RF Detector

Another important issue for high sensitivity is efficient down-conversion of receive signal to baseband domain. Figure 5.8 depicts demodulated RMS voltage of envelope detector output V_{BB} versus RF input voltage V_{RF}, whereupon the 868 MHz carrier is modulated with OOK at 100 kbit/s. The first harmonic of the baseband signal V_{BB} is measured via a spectrum analyzer and a high impedance buffer amplifier is used for proper isolation. Characteristics of simulation and measurement result for the proposed weak-inversion PMOS detector match quite good. At

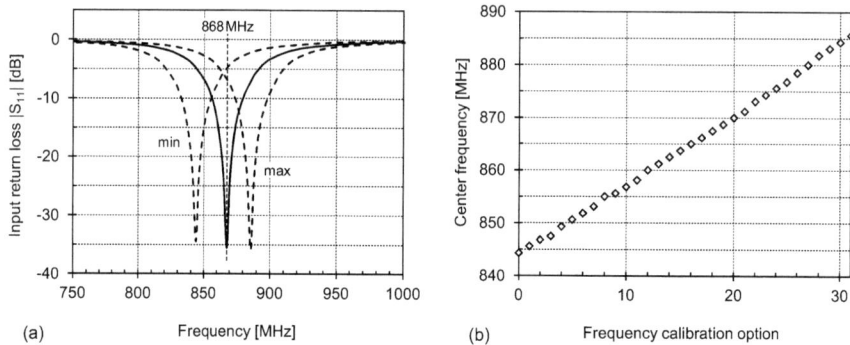

Figure 5.7: (a) RF input return loss of tunable matching network for minimum, nominal and maximum resonance frequency and (b), measured resonance frequency versus frequency calibration option

high signal strength, nonlinear effects of overdriving saturate the output magnitude, while in small signal region, square-law leads to a slope of 2 in the chart with double-logarithmic scale. As predicted from theory in chapter 3, the measured characteristic of the HSMS286 Schottky diode detector demonstrates nearly equal detector sensitivity when compared to the MOS implementation for the same bias current of 1 µA.

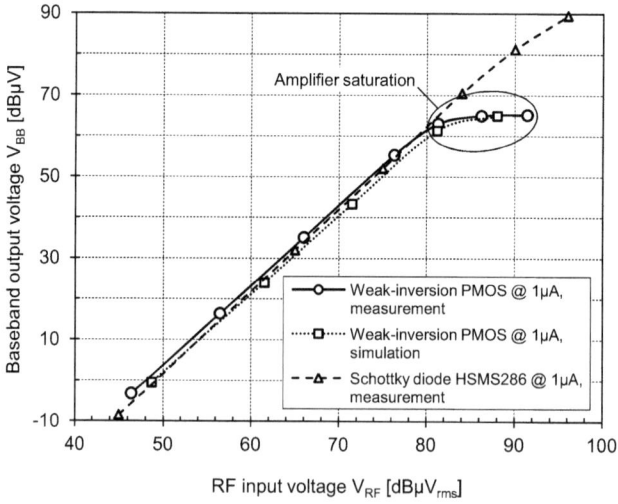

Figure 5.8: Down-conversion characteristic of PMOS detector implementation for OOK modulated RF input voltage

Preamplifier

The preamplifier ensures RF envelope detection and low noise amplification of demodulated signal coevally. Figure 5.9 compares simulated frequency response of baseband gain with measurement results for either a small on-chip blocking capacitor $C_{block} = 400\,\text{pF}$, or an additional off-chip device of $10\,\text{nF}$ in order to force low frequency cutoff. The dashed lines represent configuration options for bandwidth to provide optimal tradeoff between gain of desired signal and suppression of flicker noise. Especially at frequencies below some $10\,\text{kHz}$, measured gain is significantly below expected simulation result. The most probable explanation therefore is inaccurate bias current as a consequence from gate offset voltage between the two amplifier stages. In contrast, gain of the PGAs matches very well to simulation because of their feedback loop and hence, compensation of inexactness.

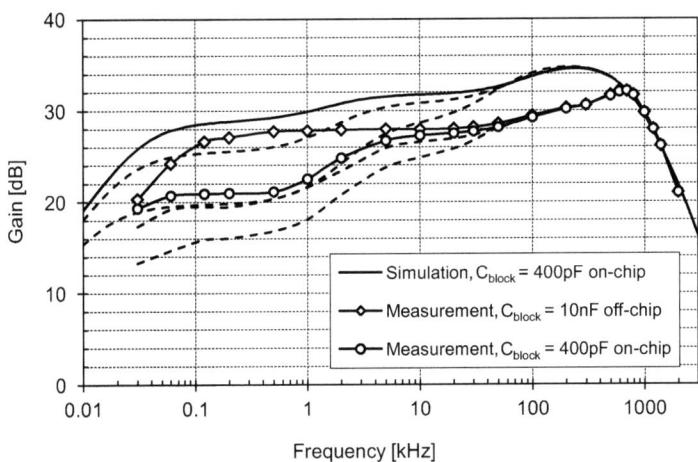

Figure 5.9: Frequency response of preamplifier gain with different options for bandwidth: Comparison of measurement with simulation

Amplifier Chain

The baseband amplifier chain conditions magnitude and bandwidth of the desired signal optimally for analog-to-digital conversion or postprocessing via correlation. Therefore, its gain is configured preferably high, but without signal clipping, and its filter bandwidth is adjusted for flicker noise suppression and anti-aliasing. Figure 5.10 depicts frequency response of the total amplifier chain that contains preamplifier, two PGAs, and a low pass filter. The measured characteristic has a maximum gain of 78 dB and a 3 dB-bandwidth of about 600 kHz. Deviation from simulation is a consequence of the previously discussed preamplifier characteristic from figure 5.9 and is acceptable without risk for difficulties.

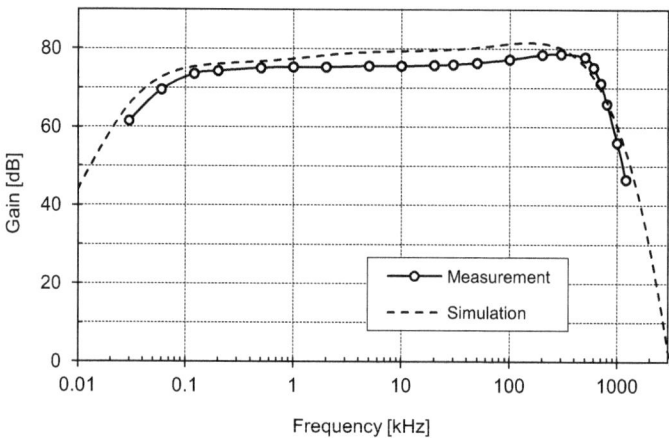

Figure 5.10: Frequency response of baseband amplifier chain for maximum gain and maximum bandwidth at 25 °C

Figure 5.11 shows measured frequency response of the implemented second-order low pass filter. Manifold tuning options for different cutoff frequencies offer 3 dB-bandwidths between 25 kHz and 560 kHz. This large range supports flexible sample rates from about 50 kS/s up to 1.1 MS/s without aliasing effect.

The input voltage referred spectral noise density in figure 5.12 implies the total noise of the baseband amplifier chain at maximum gain configuration. The graph shows noise characteristic with and without off-chip blocking capacitor for the RF detector. When decoupled externally with $C_{block} = 10$ nF, flicker noise below 10 kHz is reduced, but with the drawback of long settling time at power-up. In both cases, measured flicker noise is slightly lower than expected from simulation. In contrast, thermal noise level matches to simulated characteristics accurately. Its lower physical boundary is -33.0 dBμV/$\sqrt{\text{Hz}}$ and calculated in equation 4.3. Implemented design exceeds this limit by just 1 dB, so further reduction of noise level is possible only with higher bias current and thus increased power consumption.

Temperature characteristic of noise power is proportional to absolute temperature (PTAT), because amplifier gain is temperature stable and bias current I_{DET} in equation 4.3 also has PTAT dependency.

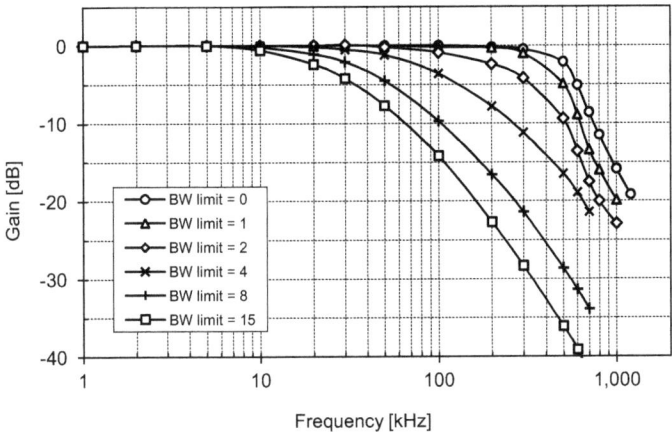

Figure 5.11: Measured frequency response of second-order low pass filter for different bandwidth configurations at 25 °C

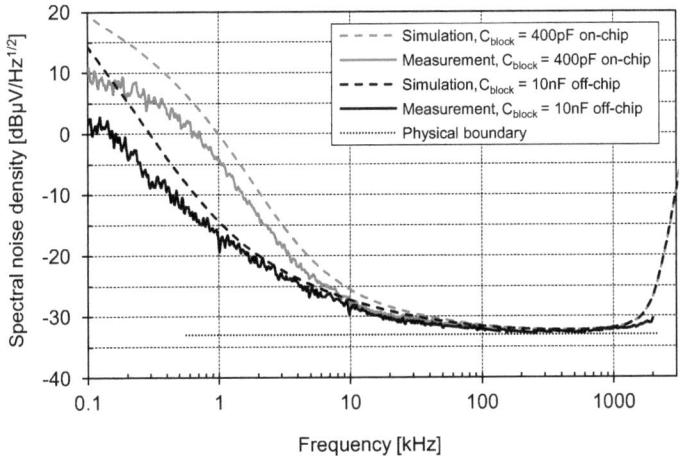

Figure 5.12: Input-referred spectral noise density of total baseband amplifier chain

Voltage Reference

The proposed ultra-low power voltage reference is designed close to CMOS technology's limitations when considering offset voltage from imperfect device matching and leakage current at elevated temperature. Consequently, several dies have been characterized in order to prove also device-to-device variation. Temperature characteristic of 8 dies is measured via a TEM in

Measurement Results and Discussion

a range from -20 °C – 100 °C and presented in figure 5.13. In opposite to simulation, the linear temperature coefficient (TC) is not compensated fully. The reason for that is most likely inaccurate simulation models of the MOS transistors, because each one is operated in subthreshold region with bias current of only 2.0 nA. This is very unusual in common designs, so accuracy of models is not optimized for such use. A worst case TC of 162 ppm/°C is still acceptable. Spread of device variation for these 8 dies is 12 mV or equivalently ±0.97 % from a nominal reference voltage of $V_{REF} = 616$ mV and without any calibration. Three die pairs have nearly equal reference voltage, so it may be concluded, that device variation relies on different die positions on the wafer, whereupon adjacent dies behave very similar and device match of separated dies is worse. Improved layout with higher grade of symmetry may reduce variation in future. Nevertheless, fast startup time below 20 μs and the small semiconductor area of 0.005 mm^2 favor this novel voltage reference as adequate candidate especially for ultra-low power voltage regulators. Slight penalties in performance and accuracy are typical for low power circuits since current consumption is only 23.4 nA at 25 °C.

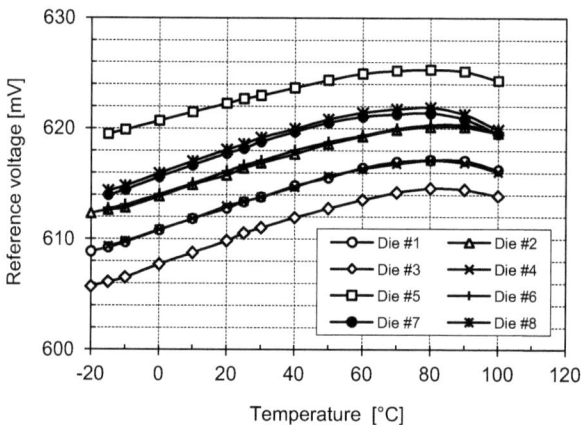

Figure 5.13: Measured temperature characteristic and device variation of novel ultra-low power voltage reference with current consumption of 23.4 nA

Figure 5.14 shows power supply rejection of voltage reference at 25 °C. Thanks to the symmetric structure of the beta-multiplier, supply voltage of up to 3.6 V causes less than 0.5 mV drift of reference voltage, so power supply rejection ratio $PSRR = 75$ dB. Furthermore, the proposed reference supports low voltage operation down to 700 mV, what is already close to the actual output of $V_{REF} = 613$ mV.

Off-the-shelf low drop output voltage regulators (LDOs) typically consume at least 0.8 μA of quiescent current such as XC6215 [31]. Therefore they are "inefficient" at loads below a few

Measurement Results and Discussion

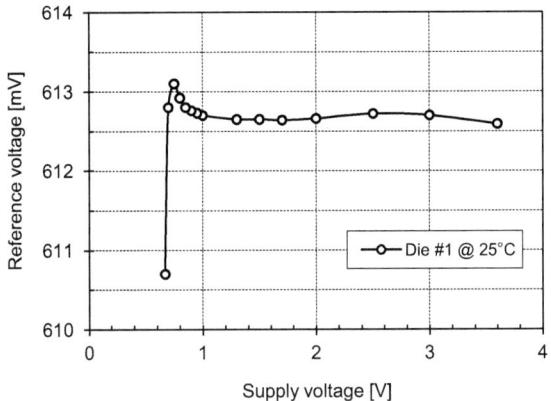

Figure 5.14: Power supply rejection of proposed ultra-low power voltage reference at 25 °C

microampere. On the other hand, they react very slowly on load steps and have settling times up to milliseconds [31]. Figure 5.15 depicts measurement result of own-current consumption for the proposed LDO with load-dependent quiescent current. This ASIC is realized in 250 nm CMOS, delivers nominally 3.0 V, and already includes the ultra-low power voltage reference. The chart presents normalized quiescent current $I_{q,N} = \frac{I_q}{I_{load}}$ versus load current I_{load} for normal linear operation and also in saturation region, where supply voltage $V_{DD} < 3.0$ V. In both cases $I_{q,N}$ is below some 10 % over a large load range. Hence, this LDO is comparatively efficient and fast at the same time, because at high load current, I_q is increased simultaneously and that raises bandwidth and speed. Current efficiency suffers only for $I_{load} < 0.1$ µA due to the static current demand of voltage reference and control loop. The minimum quiescent current is 34 nA in contrast to 800 nA for the XC6215.

5.1.3 Backend

Correlation Unit

The analog/mixed-signal correlation unit achieves a measured coding gain of up to 33 dB, but receive sensitivity of WuR is boosted only about the half value of 16.5 dB because of square-law from envelope detection. Anyhow, this benefit forces application a lot.

Figure 5.16 illustrates correlator performance in time domain. The noisy baseband signal at the correlator input results from low receive signal strength of -71 dBm. Its desired signal component from accordant digital bit pattern that is depicted above, is no more observable.

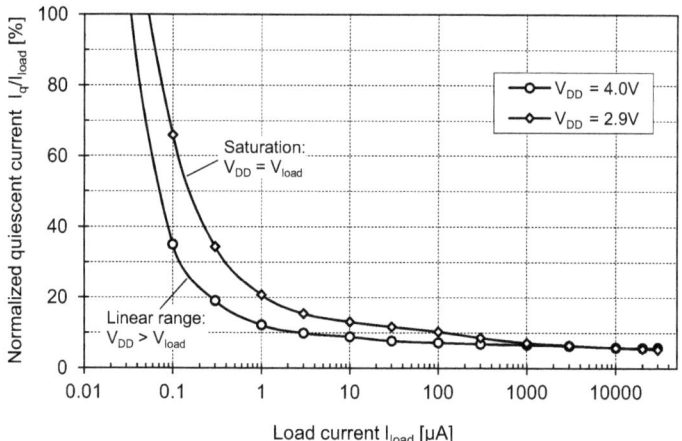

Figure 5.15: Own-current consumption of proposed ultra-low power low drop output voltage regulator (LDO) with adaptive quiescent current: Current consumption of the required voltage reference is already included.

After correlation, the WuR is still able to process this signal and successfully detect a wake-up event with probability of 99 %. When compared with simulation from figure 4.21, the measured coding gain of 33 dB is slightly below the simulated gain of 34.8 dB. The reason is most likely offset voltage of the comparators within the 128 correlation stages. For an input clock rate of $f_{CLK} = 200$ kHz, the time constant of the switched-capacitor low pass filters is evaluated to 7.2 ms, and power consumption of the complete correlation unit is 490 nW. This matches to expectations from simulation with usual uncertainty.

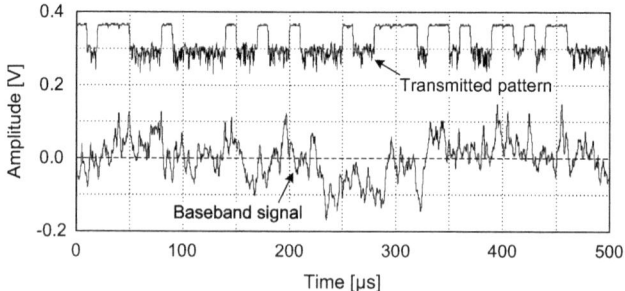

Figure 5.16: Magnitude characteristic of OOK modulated RF carrier and corresponding baseband signal for -71 dBm receive signal strength: The correlation unit is able to process this signal properly for adequate WuR functionality.

A 7-bit DAC is used to generate the decision threshold for evaluation of analog correlation result. Its resolution is 4.9 mV and the measured conversion characteristic is shown in figure 5.17. The trendline demonstrates excellent linearity. However, around zero point low offset voltage of 5 mV is recognized. This is an effect of the DAC's buffer amplifier that is used to drive high capacitive loads. Offset is not relevant for WuR operation, since only strict monotonicity is of importance and absolute accuracy of output value is of minor interest. Measurements indicate that even higher resolution would be possible via this basic DAC approach with capacitive voltage dividers.

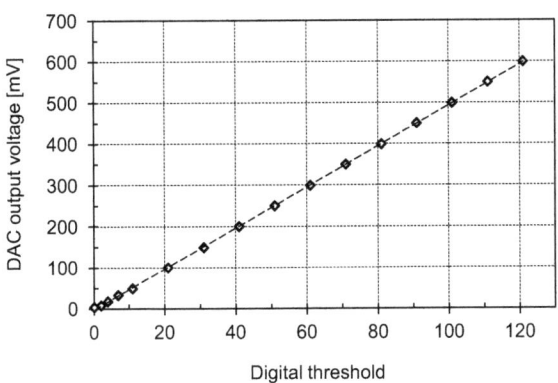

Figure 5.17: Conversion characteristic of 7-bit digital-to-analog converter for generation of adjustable decision threshold within mixed-signal correlation unit

For comparison, power consumption of a conventional correlation scheme in digital domain is simulated. Also if consumption of analog-to-digital conversion is excluded, a simplified digital correlation scheme would consume at least 6.2 µA at 1.0 V supply. The basis for this layout-extracted and already optimistic power simulation is a 128-stage correlation chain with digital standard library elements. It contains of a 128 × 3 bit shift register for data storage, all in all 501 bit of full adders for hierarchical summation, and a 11 bit comparator for evaluation of result. The feature for correlation over repeated code sequences is omitted for simplicity. When compared with mixed-signal correlation unit, power consumption of the digital implementation would be at least 13 times higher and hence, unsuitable for the intended application.

A/D Converter

In order to support off-chip digital signal processing, two types of analog-to-digital converters (ADCs) are implemented. Characterization of the 3-bit SAR ADC is done by means of a slowly rising triangular input signal of 100 Hz. Each of the 7 quantization thresholds is then monitored simply via a changeover from one code to the next. Figure 5.18 depicts the resulting transfer characteristic. The adjustable reference voltage of $V_{REF} = 628\,\text{mV}$ equals the highest quantization threshold. Linearity of the ADC is quite good in spite of the power-optimized design with current consumption of only 145 nA at 100 kS/s. The small offset voltage of $V_{O,ADC} = -9\,\text{mV}$ is a consequence of comparator offset and is irrelevant for AC-coupled operation.

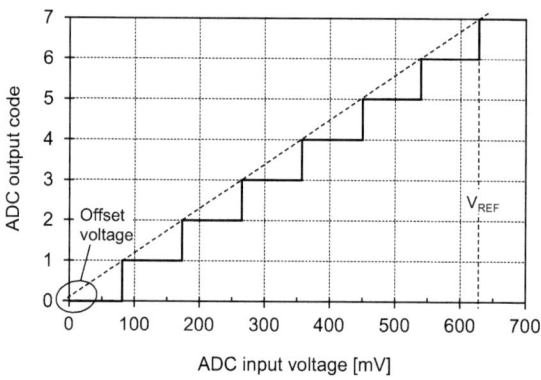

Figure 5.18: Transfer characteristic of ultra-low power successive approximation register (SAR) ADC with resolution of 3 bit: Reference voltage $V_{REF} = 628\,\text{mV}$

The alternative ADC implementation with PWM output has inherently high linearity by design. It consumes 161 nA at 100 kS/s and the reference pulse width can be selected between 0.42 μs and 13.2 μs. This flexibility allows a large range of sample rates from 75 kS/s up to 2.5 MS/s at high resolution.

5.1.4 Wake-up Receiver System

For test and characterization of the complete WuR system, the ASIC dies are mounted onto PCB modules that include all necessary auxiliaries for voltage supply and clock generation in order to enable field trials with low effort. The used antennas are quarter-wave monopoles with high radiation efficiency and are matched to 50 Ω at 868 MHz. They comprise a structure to emulate ground plane for well-defined radiation footprints (see figure 2.5).

Micro photographs of the ASIC dies are presented in figure 5.19. The passivation layer on top of the chip contains a polyimide coating and is widely transparent, so the last metal layer (aluminium) and its fill structures are visible. Near the RF-input pad, three metal-insulator-metal coupling capacitors and parts of the surrounding pad-ring on layers $Metal4 - Metal6$

are observable through fill structures. To minimize parasitic effects of packaging and negative impact to RF performance, the dies are bonded directly to gold-plated FR4 boards.

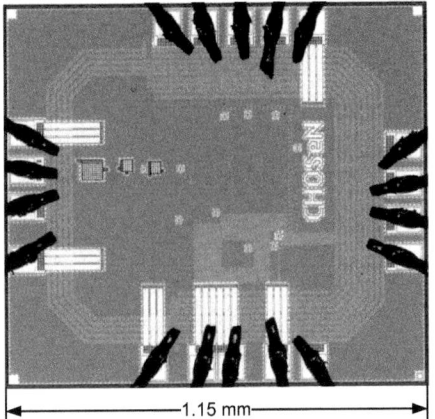 

Figure 5.19: Photos of WuR chip: The die is glued and bonded directly to a gold-plated FR4 PCB via aluminium wires of 20 μm diameter.

In order to provide high flexibility for application, the wake-up receiver can operate in one of three power states. This is illustrated in figure 5.20 together with typical transition times. In active state, the WuR is fully powered, listens to incoming wake-up calls and consumes 2.4 μW in total. In power-save state, only configuration registers and SPI are supplied for content retention and fast resume of operation. Here, power consumption is only 12 nW. Power-down state in contrast, shuts down the complete WuR and this results in lowest power consumption of 11 nW. The additional saving compared with power-save mode is low, but especially at elevated temperature, the residual static leakage current of the ESD structures and the configuration registers becomes dominant. So power-down state then is beneficial. Basically transition times between power states are short. An exception is activation of RF listening mode. The reason for that is the long settling time of the baseband amplifier because of its low cutoff frequency.

Noise-caused generation of wake-up events without desired receive signal leads to unnecessary activation of the main receiver and hence, waste of energy. So this false wake-up rate (FWR) has to be sufficiently low to avoid significant loss of battery lifetime. The slicer evaluates the analog correlation result and generates wake-up events if its decision threshold is exceeded. Figure 5.21 shows measured FWR versus adjustable wake-up threshold. Omnipresent noise can trigger a wake-up event, so already a slightly increased decision threshold pushes down FWR a lot. This steeply characteristic is a consequence of the Gaussian distribution of residual

Measurement Results and Discussion

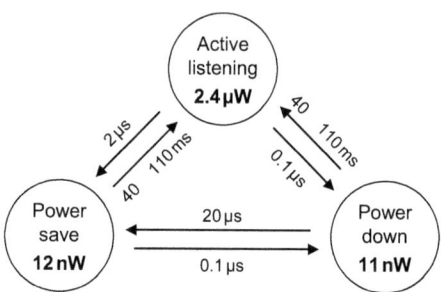

Figure 5.20: WuR power states and transitions with measured power consumption at 25 °C

noise.

For performance measurements with the WuR system, the decision threshold is configured at 27 mV for $FWR < 10^{-3}/s$ or equivalently one unwanted wake-up interrupt every 16 minutes at most. Even higher threshold would increase robustness, but simultaneously decrease sensitivity.

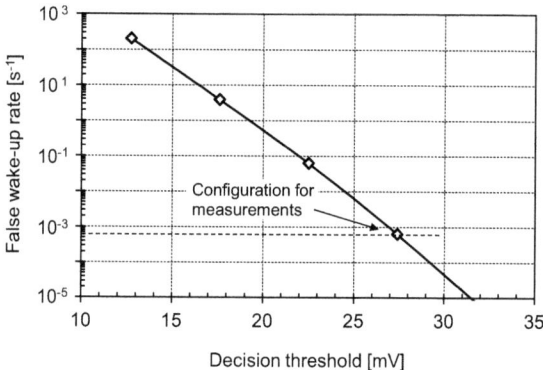

Figure 5.21: Dependency of false wake-up rate (FWR) from adjustable decision threshold: The slicer of the correlation unit can generate unwanted wake-up events because of omnipresent noise.

Beside ultra-low power consumption, the most important key parameters of WuRs are receive sensitivity and latency. Performance of the implemented design is illustrated in figure 5.22. The chart shows probability for missed detection of wake-up calls and also corresponding latency of wake-up event detection versus receive signal strength. With decreasing RF input power, error probability for wake-up detection grows rapidly. One can say that there is a sensitivity boundary for proper reception of wake-up events. If probability for successful detection should be at least $P_{DET} = 1 - \overline{P_{DET}} = 99\,\%$, then receive sensitivity of $S_{99\,\%} = -71\,\text{dBm}$ is obtained. The second curve shows the accordant mean wake-up latency of 5 ms for wake-up interrupt generation. It decreases below 1 ms with increasing RF input power thanks to the inherent linearity of the correlation concept with periodic code sequences. This way, the benefit of high signal strength is utilized to reduce delay and responsiveness of the wireless node is forced.

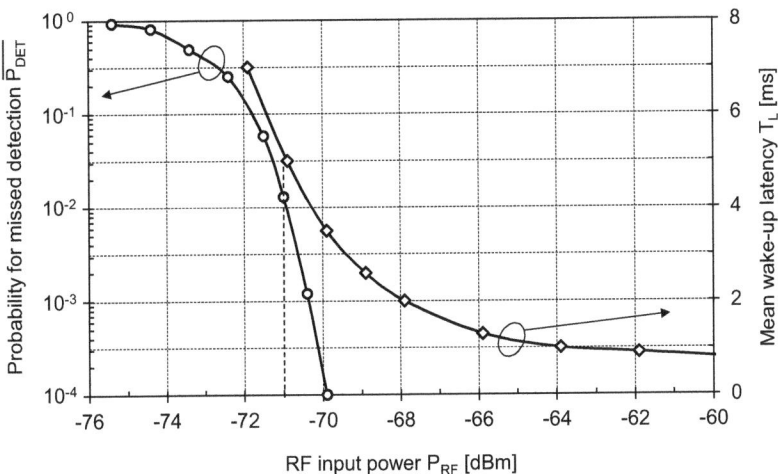

Figure 5.22: Impact of RF signal strength to probability and latency of wake-up event detection: Insertion loss of the off-chip SAW filter is already included.

Additionally, measured results for receive sensitivity are proven via field trials in real environment. Therefore, the WuR module and the transmitter module are mounted onto masts for outdoor measurements in order to ensure proper free-space conditions with free Fresnel ellipsoid. A radio link distance of up to 304 m (GPS coordinates WuR: 48°12′11.94″N, 15°33′16.04″E, transmitter: 48°12′19.35″N, 15°33′25.76″E) is achieved with 6.4 dBm transmit power, and quarter-wave monopole antennas with gain of 1.0 dBi. The WuR configuration is adjusted for a false wake-up rate of 5×10^{-4}/s, 99 % detection probability, and a correlation length of 7 ms. Transmit signal at 868 MHz has a bit rate of 100 kbit/s and OOK modulation. Measurement result confirms with expectation from theory, when link budget is calculated with equation 2.1.

Indoor radio links typically reach up to $\approx$10 m, if two walls of concrete are considered in between. This performance emphasizes practical usability for short-range communication, especially for ultra-low power systems where the WuR's consumption of 2.4 µW is in the same range when compared to demand of microcontrollers in deep-sleep mode with enabled realtime clock.

5.2 Comparison

Comparison of measurement results with simulation and theoretic background is necessary to prove the proposed concept, evaluate the implemented design, and gain further knowledge for refinements. This is all the more important, if some unusual method of operation for simulation models of transistors is deployed.

Temperature characteristic of current reference and proposed ultra-low power voltage reference from figure 5.23 is analyzed and discussed. The chart illustrates measured characteristics via solid lines, accordant simulation outcome via dashed lines, and trendlines from theory start at the absolute zero-point of 0 K. Since reference current determines biasing for the complete ASIC, a good match to simulation promises proper operation. The upper curves show the PTAT voltage characteristic V_{PTAT} of the current reference's beta-multiplier structure from figure 4.13. Its calculated value from equation 4.2 is 53.4 mV at 25 °C. Measured data is 53.7 mV and simulation yields 53.8 mV, so the implemented design seems to be robust because of excellently fitting values. The central set of lines depict the temperature characteristic of the reference current generator. Calculation for 25 °C results in $I_{REF} = 49.5$ nA, measurement produces 49.4 nA, and simulation data is 49.65 nA. These marginal deviations are somewhat surprising since variations in fabrication process can yield to a worst case tolerance of 20 % for the current determining polysilicon resistor R. Anyhow, at least V_{PTAT} is defined by physical constants in equation 4.2, so its accuracy is usually guaranteed by design. Increasing deviation of reference current from ideal PTAT characteristic at high temperature is the consequence of leakage current and may be a result of current measurement procedure that uses auxiliary circuitry. The lower graph presents the temperature characteristic of supply current of the ultra-low power voltage reference. In spite of the extremely low bias current of 2 nA per transistor, measurement confirms with simulation also at high temperature.

In figure 5.24, overall performance of the proposed WuR concept is compared with designs from state-of-the-art literature. The graph depicts the tradeoff between power consumption and receive sensitivity of dedicated WuRs for WSNs. It includes data of figure 2.19 from related work as well as performance of the actual design. The realized WuR concept with sensitivity of -71 dBm explicitly has lowest power consumption of only 2.4 µW. The best performing competitor from [SDM09, SD09] demands more than 5 times higher power and has much less sensitivity of -57 dBm. When compared with performance of physical implemented designs with at least equal sensitivity from [PRG09, HRW+10], the realized WuR beats power consumption of these designs by more than one order of magnitude. If significantly increased receive sensitivity of around -90 dBm is required, state-of-the-art work from [OCR05, CBM+06] already consume at least 100 times more than the actual design.

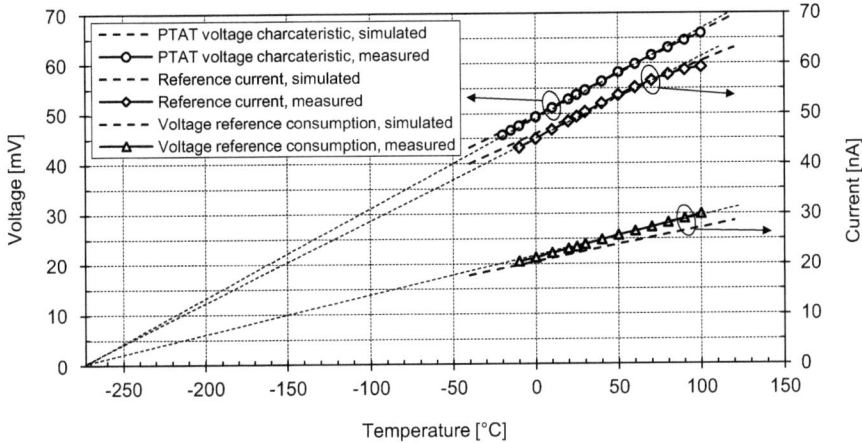

Figure 5.23: Temperature characteristic of reference circuits: Comparison of simulation and measurement of PTAT voltage and corresponding reference current as well as supply current of ultra-low power voltage reference.
PTAT: proportional to absolute temperature

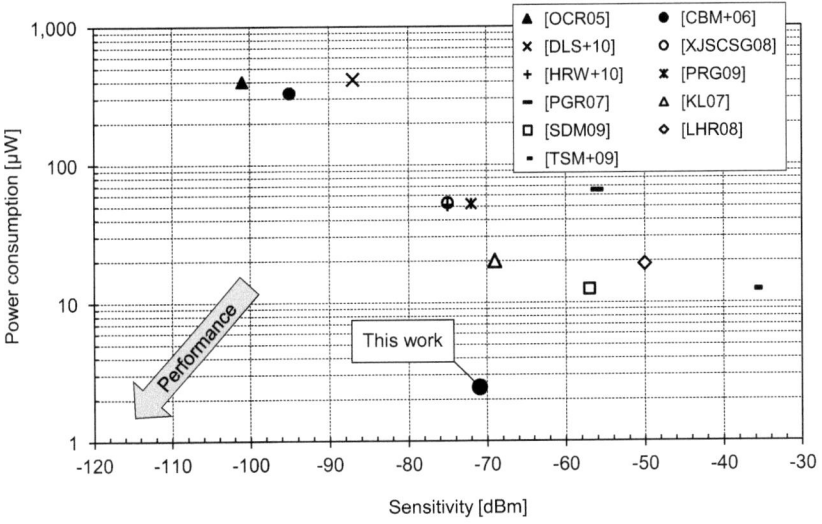

Figure 5.24: Performance comparison of actual WuR design with state-of-the-art

One can consider an imaginary line through data points of best performing related work. Then, the realized design clearly outperforms state-of-the-art. The main reason for that is the innovative combination of a low-power envelope detector frontend with mixed-signal correlation technique in baseband domain. The latter exploits high coding gain for enhanced sensitivity at concurrently low power consumption. Furthermore, the whole ASIC implementation is close to the physical boundaries of the proposed architecture. So there is minor margin for further reduction of power consumption without compromise in sensitivity or latency.

5.3 Performance Summary

The realized design of an ultra-low power wake-up receiver for WSN comprises a companion chip solution to extend functionality of conventional sensor nodes. Via wake-up interrupt and SPI interface for microcontroller communication, the implemented ASIC allows for ordinary integration into a targeted application.

The innovative concept with combination of a simplified low-power radio frontend and an analog correlation technique for signal postprocessing in low frequency domain brings major advantages in receive sensitivity when compared to prior art and was also presented at ISCAS 2011 [HMH11].

Table 5.1 presents typical operating conditions for the actual WuR design. It specifies voltage ranges and accordant current demand for all supply domains. A large range of bit rates and clock frequencies is supported thanks to flexible configuration options for gain and signal bandwidths. The clock accuracy must be better than 1000 ppm to guarantee maximum coding gain from the correlation unit. Specification of digital I/O interfaces and chip initialization procedure are commonly.

Table 5.2 summarizes most important performance parameters and ASIC characteristics of the realized WuR solution for 868 MHz. A receive sensitivity of -71 dBm is achieved for a wake-up event detection probability of 99 % and a false wake-up rate of 10^{-3}/s. Thereby, loss of the off-chip SAW filter is already included. The comparatively high sensitivity is a consequence of the low-power mixed-signal correlation unit in baseband domain that operates with circular 64 bit code sequences, variable correlation length of 300 bit or 700 bit, and supports modulation bit rates from 10 kbit/s – 200 kbit/s. RF input impedance has very low loss in radio listening mode for adequate voltage transformation by the matching network. Measured supply currents of essential WuR parts fit well to the simulation results and confirm the total consumption of 2.4 µW as expected. Successful ASIC operation is tested over a temperature range from -20 °C to 100 °C and that indicates robust design implementation. When bonding pad structures are included, the WuR's chip area is 1.15 mm². Thereof, actual core functionality occupies only 0.26 mm².

Symbol	Parameter	Conditions	Min	Typ	Max	Unit
V_{DDD}	Digital core supply voltage		1.0		1.2	V
V_{DDA}	Analog core supply voltage		1.0		1.2	V
V_{DDIO}	I/O supply voltage		V_{DDD}		3.6	V
I_{DDD}	Supply current for V_{DDD}	$V_{DDD} = 1.0\,V$, active listening mode		0.5	1.0	µA
I_{DDA}	Supply current for V_{DDA}	$V_{DDA} = 1.0\,V$, active listening mode		1.85	2.4	µA
I_{DDIO}	Supply current for V_{DDIO}	$V_{DDIO} = 3.3\,V$, no load driven		0.1		µA
f_{CLK}	Baseband sample rate	Duty-cycle = 50/50	20	100	400	kHz
	Frequency accuracy	Maximum sensitivity			1000	ppm
f_{SCK}	SPI clock frequency	$V_{DDIO} = 3.3\,V$	DC		20	MHz
$t_{r,f}$	Digital input rise/fall time	$V_{DDIO} = 3.3\,V$			100	ns
t_{reset}	Register reset duration	After rising edge of EN input			2.0	µs

Table 5.1: Operating conditions for realized WuR ASIC at 25 °C

Parameter	Conditions	Value	Unit
Wake-up/receive sensitivity	7 ms correlation period with mixed-signal correlator, 100 kbit/s OOK modulation, 2.2 dB insertion loss of SAW filter included, $P_{DET} = 99\,\%$, FWR = 10^{-3}/s	-71	dBm
Carrier frequency	Defined by off-chip SAW filter and matching network	868	MHz
Raw bit rate	Mixed-signal correlator limit	10 – 200	kbit/s
Wake-up pattern/ address length	Mixed-signal correlator	64	bit
Correlation length	Correlator options	300/700	bit
RF input impedance	ASIC without matching at 868 MHz, RF switch off	8.5 - j318	Ω
	ASIC without matching at 868 MHz, RF switch on	118 - j45	
Supply current	Total WuR, active listening, $V_{DD} = 1.0\,V$	2.4	µA
	Frontend active, $V_{DDA} = 1.0\,V$	1.85	
	Correlator at 100 kbit/s, $V_{DDD} = 1.0\,V$	0.49	
	Ultra-low power 615 mV reference	23	nA
	WuR in power-save mode, $V_{DDD} = 1.0\,V$	12	
	WuR in power-down mode, $V_{DDD} = 1.0\,V$	11	
Power-up time	Dependent on filter configuration	40 – 110	ms
Temperature range	Operation tested	-20 – 100	°C
Chip area	Pad ring included, 130 nm CMOS	1.15 × 1.0	mm²

Table 5.2: Typical wake-up receiver key parameters measured at 25 °C

With the help of theory, a sensitivity boundary for the chosen WuR concept can be calculated/ estimated for fixed key parameters such as available power consumption and allowed wake-up latency. It depends from adequate RF impedance transformation with minimized loss of the

Measurement Results and Discussion

matching network for a high RF voltage transformation ratio, maximum nonlinear characteristic of the detector device for high down-conversion efficiency, minimum baseband noise level, and latency-limited correlation length. All these conditions have been reached closely with the realized ASIC design. Consequently, there is minor margin for further improvement of receive sensitivity for the proposed architecture at similar level of power consumption.

When compared to expectations from simulation (see table 4.3), the most important parameters of power consumption, latency, and sensitivity have been met. To the best of the author's knowledge, this is the only realized wake-up receiver design with power consumption in the range of a few microwatts and comparatively high sensitivity. It provides a mostly full integrated add-on solution for conventional wireless sensor nodes.

6 Conclusion and Outlook

Penetration of human's daily life with WSNs is growing with enormous speed. Many users do not notice when they profit from wireless sensor nodes in their vehicle, in traffic infrastructure or at their working place. Well established applications are remote controls, security and safety systems in office buildings, or intelligent control for comfort and energy saving systems. In future, WSNs will capture additional application fields due to enhanced features and extended lifetime of network service.

This chapter summarizes major findings and results of this work, and gives an outlook over possible enhancements for the developed WuR solution. Furthermore, it outlines fields of interest for development and research as well as potential for future application and resulting consequences for the environment.

6.1 Summary and Conclusion

In order to bridge the gap between power consumption and network reactivity in conventional wireless sensor nodes, an ultra-low power wake-up receiver was designed, since up to now, there is no commercial product available on the market for such application. This dedicated and highly power-optimized add-on receiver can listen to the radio channel all the time and react immediately on incoming data packets with low latency. The main design intention was to reduce power consumption and enhance receive sensitivity when compared to state-of-the-art solutions from literature. This is achieved by the concept of combination of a low-power RF detector based radio frontend together with an innovative analog correlation technique in baseband domain. Consequently, elimination of power-hungry signal processing in RF domain leads to a sensitivity penalty due to the simplified frontend, but this fact is compensated at least partly via high coding gain from correlation that has concurrently very low power consumption.

Conclusion and Outlook

This work comprises development and realization of a novel concept for ultra-low power wake-up receiver designs for wireless sensor nodes. It was implemented in an ASIC solution for 868 MHz carrier frequency. The realized 130 nm CMOS chip includes an RF envelope detector, a low noise baseband amplifier, a PGA, a novel analog/mixed-signal correlation unit, as well as auxiliaries for chip characterization and stand-alone operation. A receive sensitivity of -71 dBm is achieved for 99 % detection probability and at a noise-caused false wake-up rate of 10^{-3}/s. The total power consumption is only 2.4 µW from 1.0 V supply. The main reason for that is a baseband signal correlation technique with 64 bit code sequences over up to 7 ms. Off-chip components are required only for radio frequency band selection and impedance matching.

The main differences and most important enhancements of the actual WuR design over prior art are summarized in the following items:

- Correlation approach: Thanks to the architecture with signal postprocessing via correlation principle, the desired input signal is accumulated over long periods and noise bandwidth is reduced to a minimum. So the tradeoff between receive sensitivity and latency can be adjusted according to the actual requirements. The code sequences for correlation represent individual WuR address information at the same time. This way, node access is handled inherently by the correlation unit, and a separate address decoding stage becomes obsolete.

- High quality circuit design: Many circuit implementations from literature suffer from a bottleneck due to improper design of unexperienced developers. To cope with this problem, theoretic boundaries for the proposed concept are calculated first. The final circuit implementation reaches them almost with minor margin for further improvements. Therefore, specific and optimized solutions for subsystems are realized via sophisticated full-custom design. The benefit is comparable performance, but at much reduced power consumption than state-of-the-art.

- High receive sensitivity: Thanks to the correlation approach, signal-to-noise ratio in baseband domain is increased by large coding gain and accordingly, sensitivity of square-law detector frontend is enhanced too.

- Power consumption: Because of rigorous low-power circuit design and especially due to the innovative analog correlation technique that is based on a switched-capacitor principle, power consumption of the total WuR is reduced to 2.4 µW in listening mode. When compared to state-of-the-art designs with comparably high receive sensitivity, the achieved power consumption beats competitors by an order of magnitude (see figure 5.24).

- Integration density: An integrated RF detector, a low number of off-chip devices, compatibility to standard CMOS technology, and low space requirement of $550 \times 480\,\mu m^2$ for the actual WuR core yield to the potential for low cost integration and high volume production. The designed WuR ASIC represents a complete add-on solution to common WSN nodes.

The first silicon of the developed WuR ASIC is already fully functional. Measurement results of prototypes confirm with expectations from simulation in a large extent. Operation has been tested and demonstrated successfully with the help of field trials.

Figure 6.1 shows the assembly of a WuR prototype module as well as a photo of the 200 mm wafer that was fabricated via shard reticle mask set and contains around 110 WuR dies. The large number of surface-mounted components provides extensive options for test and ASIC characterization. The SMA connector on the left hand side is an analog test interface, while the opposing $50\,\Omega$ connector is the actual antenna input.

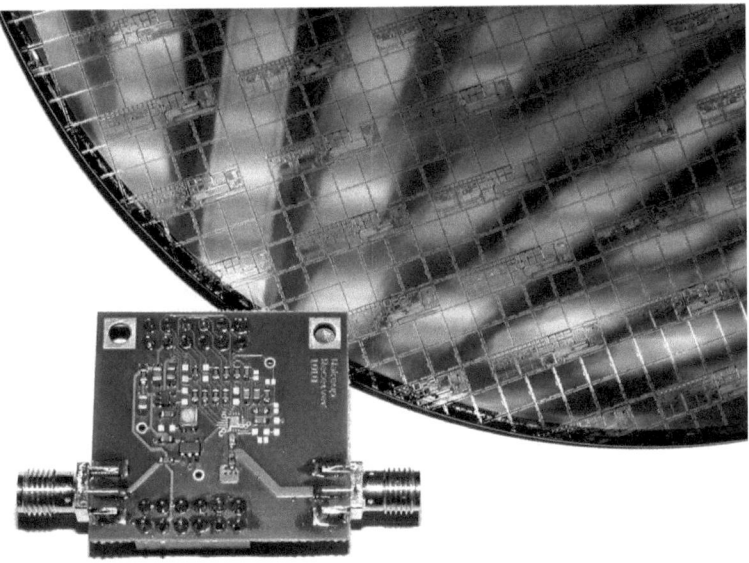

Figure 6.1: Assembly of WuR module with PCB bonded die and corresponding 200 mm prototype wafer, fabricated via shared reticle mask set in 130 nm CMOS technology

The small modules from figure 6.2 represent a very basic demonstrator with wake-up functionality. Each node is button-cell powered, assembled onto a PCB of $18 \times 18\,mm^2$, and equipped with shorted monopole antennas for 868 MHz. On button-press, the transmitter module flashes

Conclusion and Outlook

its LED and sends a single wake-up packet. The WuR module in contrast listens for incoming packets all the time, and flashes its own LED as soon as it detects a wake-up event. In idle mode, the transmitter module draws 0.5 µA from a 3.0 V lithium battery and the WuR module's current demand is 3.2 µA from an 1.5 V silver-oxide battery, if current demand for LEDs is excluded. Thereof, 0.9 µA is consumed by the 32 kHz oscillator, the microcontroller and the external 1.0 V voltage regulator. Capacity of the transmitters's battery is nominally 225 mAh and respectively 340 mAh for the receiver module. So the expected lifetimes are more than 10 years, if LED consumption is excluded and the mean packet rate is no more than 6 per hour. With setup and conditions from table 5.2 and quarter-wave monopole antennas from figure 2.5, up to 300 m of radio link distance were measured under free-space condition.

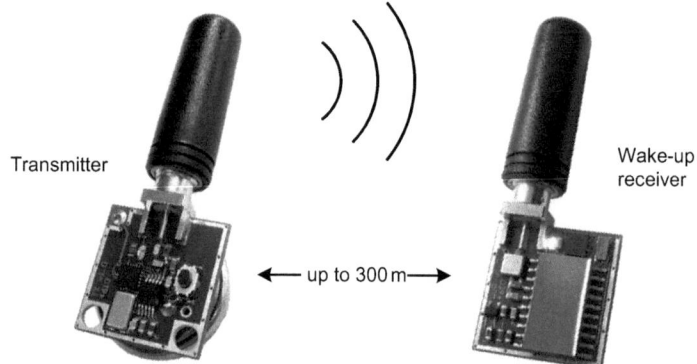

Figure 6.2: Photo of WuR demonstrator: Both transmitter module and the WuR module are full-functional button-cell powered wireless nodes with lifetimes of typically more than 10 years. They are assembled on an $18 \times 18\,\text{mm}^2$ PCB and operate with shorted monopole antennas at 868 MHz. For demonstration, the WuR die is glued and bonded into a SO20 ceramic package.

The targeted performance for an appropriate wake-up receiver solution was power consumption below 5 µW, worst-case latency of 10 ms, and receive sensitivity of at least -70 dBm. The implemented ASIC design reaches all goals and partly has even better performance than expected. When compared with state-of-the-art work from literature, it can be concluded that the realized WuR concept has mainly reduced power consumption. The benefit over prior art is more than one order of magnitude (see figure 5.24), if competing designs with equal receive sensitivity are considered. This clearly emphasizes relevance of the actual work and allows new research fields for ultra-low power applications, where the WuR's power consumption is in the same range compared to deep-sleep modes of microcontrollers with enabled realtime clock. Thereby, the achieved sensitivity is still sufficient at least for wireless communication over short range.

6.2 Enhancements and Future Work

So as to extend functionality of the realized wake-up receiver further, the ASIC can be enhanced by additional functions and configuration options. These can support cost-efficient integration into wireless nodes, provide increased flexibility for a broadened range of applications, or even gain performance for special aspects via adjustable tradeoffs. Possible enhancements and future working fields for the actual WuR design are:

- BAW resonator frontend: Currently, a SAW filter is used for off-chip RF channel selection. It may be replaced by a bulk acoustic wave (BAW) resonator such as in [PGR07, PRG09]. The benefits would be increased integration density with potential for cost reduction, sharp frequency selection, possibility for frequency tuning, and integration of currently off-chip power matching network into the resonator circuit.

- Adoption of carrier frequency: The RF frontend may be redesigned for operating frequencies between 100 MHz and 3 GHz according to requirements of the target application.

- LNA option: In order to enhance receive sensitivity, an additional and duty-cycled low noise amplifier may be applied to gain RF input signal with the drawback of significantly increased power consumption.

- RSSI: The feature of received signal strength indication (RSSI) is beneficial for many wireless communication protocols. Is can be extracted from baseband amplifier chain and may support network operation effectively.

- On-chip LDO voltage regulator: An additional voltage regulator that contains the presented ultra-low power voltage reference allows for on-chip generation of 1.0 V supply for analog and digital core. Then the currently external voltage regulator becomes obsolete. Alternatively, a charge pump based DC/DC converter can supply the WuR core directly from I/O voltage with high efficiency, and an adjustable supply voltage for the digital domain minimizes power consumption of the signal processing unit.

- Clock management unit: A low-power fractional-N PLL can be used to operate the WuR directly with an external 32.768 kHz reference frequency, and concurrently offers high grade of flexibility concerning bit rate. The analog output signal of an already present crystal oscillator from the microcontroller's realtime clock can be utilized as clock reference. Thus, power consumption of an additional oscillator is avoided. Alternatively, a low-power RC oscillator can generate the required clock, if its frequency is calibrated periodically.

- Digital signal processing core: Optional digital implementation of an ultra-low power correlation scheme with full-custom logic design offers versatile configuration options when compared to analog correlation unit, but has the drawback of higher power consumption.

- Data reception: The current WuR architecture could be extended by a signal processing unit with clock and data recovery for basic user data reception at low bit rates. Then, the WuR is upgraded to a full-fledged receiver and may replace the dedicated main receiver.

- Duty-cycled superheterodyne approach: If major enhancement of receive sensitivity is required, the WuR's radio frontend has to use a superheterodyne principle for signal down-conversion. Via duty-cycling of local oscillator, mixer and IF amplifier in a large extent, much power can be saved. Anyhow, simulation shows that consumption of such an architecture is higher when compared with the presented WuR concept. Nevertheless, significantly increased sensitivity and enhanced robustness against interference may be preferred for custom application.

- WuR based MAC protocols: Novel WuR powered MAC protocols as well as integration of the wake-up feature into conventional communication protocols represent a large field for research activity in WSN.

Up to now, there is no commercial product of a wake-up receiver available on the market that fits to the desired applications in chapter 1. So many of the outlined working fields are of interest for research and especially for product development.

6.3 Outlook and Vision

With the help of the presented WuR performance, energy demand of wireless sensor nodes for ultra-low power application can be reduced dramatically to below $5\,\mu W$. Without the demand for scheduling MAC protocols thanks to the always-on WuR, the worst-case latency for starting-up communication is in the range of a few milliseconds. This circumstance allows realtime communication also for multi-hop connections. Consequently, the wireless network is operated without sophisticated protocols that typically require comparatively much amount of energy for node synchronization, especially in case of ad-hoc systems with frequently joining or leaving sensor nodes.

In classical metering applications, a mobile master unit requests the collected data from wireless sensor nodes, and these transmit metered values directly to the sink node. Then, immediate detection of the inquiry requires quasi realtime capability that can be provided only by a WuR.

Concurrently, ultra-low power consumption for service lifetimes of more than 10 years without battery replacement is an obvious demand. One typical application example is retrofit equipment for wireless readout of electrical meters, water flow meters, or gas meters. Lots of cost would be saved, if data is collected by a service operator via a wireless data collection node when driving past, instead of manual inspection in each household.

In building automation, a multiplicity of sensors and actuators require realtime capability too for adequate responsiveness during human interaction, but also low power consumption for maintenance-free operation with small and low-cost batteries.

Another novel application with promising future potential is structure health monitoring. Buildings and vehicles in harsh environment are monitored via a network up to thousands of sensors that detect damage of the mechanical structure. Synchronized readout of recorded data or immediate notification in case of an alarm condition require WuR functionality. In an aeronautic scenario, the energy harvester powered sensor nodes are mostly embedded into the aircraft's structure because maintenance, weight, and low reliability of cabling is no option. With the help of permanent supervision, maintenance intervals can be scheduled according to the actual demand. Furthermore, system safety is enhanced because of early detection of corrosion or material fatigue. In addition, extra safety margin of mechanical structure can be reduced without compromise in safety thanks to continuous monitoring. This decreases weight, and the benefit would be enormous saving of energy, cost, and environmental pollution already due to reduction of carbon dioxide emission, if world-wide air traffic is considered.

However, the research field on novel and upcoming wireless applications with ultra-low energy budget is growing. Easement of daily human's life is rising continuously due to increased grade of automation with ubiquitous and smart network services that are integrated into environment and infrastructure.

The consequence of versatile novel applications that can profit from wake-up receivers is a good market potential. From strongly rising research activity within the last few years, I expect that wake-up receivers will win relevance enormously also for commercial products as soon as their performance is adequate for practical use cases.

Literature

[AJK05]   H. AL-JUNAID and T. KAZMIERSKI, "Analogue and mixed-signal extension to SystemC," in *IEE Proceedings - Circuits, Devices and Systems*, Dec. 2005, pp. 682–690.

[Ann98]   A.-J. ANNEMA, "Low-Power Bandgap Feferences Featuring DTMOST's," in *Proceedings of the 24th European Solid-State Circuits Conference 1998. ESSCIRC'98*, Sep. 1998, pp. 116–119.

[Apa88]   J. APARICI, "A Wide Dynamic Range Square-Law Diode Detector," *IEEE Transactions on Instrumentation and Measurement*, vol. 37, no. 3, Sep. 1988, pp. 429–433.

[BFND06]   M. I. BROWNFIELD, A. S. FAYEZ, T. M. NELSON, and N. DAVIS, "Cross-layer Wireless Sensor Network Radio Power Management," *IEEE Wireless Communications and Networking Conference. WCNC 2006*, vol. 2, Apr. 2006, pp. 1160–1165.

[BKS+09]   T. BECKER, M. KLUGE, J. SCHALK, K. TIPLADY, C. PAGET, U. HILLERINGMANN, and T. OTTERPOHL, "Autonomous Sensor Nodes for Aircraft Structural Health Monitoring," in *IEEE Sensors Journal*, vol. 9, no. 11, Nov. 2009, pp. 1589–1595.

[CBM+06]   B. W. COOK, A. BERNY, A. MOLNAR, S. LANZISERA, and K. S. J. PISTER, "Low-Power 2.4-GHz Transceiver With Passive RX Front-End and 400-mV Supply," in *IEEE Journal of Solid-State Circuits*, vol. 41, no. 12, Dec. 2006, pp. 2757–2766.

[CG00]   S. CHOPRA and R. S. GUPTA, "Subthreshold conduction in short-channel polycrystalline-silicon thin-film transistors," in *Semiconductor Science and Technology*, vol. 15, Feb. 2000, pp. 197–202.

[CMP+06]   J. CHABLOZ, C. MÜLLER, F. PENGG, A. PEZOUS, C. ENZ, and M. DUBOIS, "A Low-Power 2.4GHz CMOS Receiver Front-End Using BAW Resonators," *Solid-State Circuits Conference - Digest of Technical Papers, 2006. ISSCC 2006. IEEE International*, San Francisco, CA, Feb. 2006, pp. 1244–1253.

Literature

[Con01]   L. CONG, "Pseudo C-2C Ladder-Based Data Converter Technique," *IEEE Transactions on Circuits and Systems II: Analog and Digital Signal Processing*, vol. 48, no. 10, Oct. 2001, pp. 927–929.

[dCFP05] L. H. DE CARVALHO FERREIRA and T. C. PIMENTA, "A CMOS Voltage Reference Based on Threshold Voltage for Ultra Low-Voltage and Ultra Low-Power," *The 17th International Conference on Microelectronics. ICM 2005*. Islamabad, Pakistan: IEEE, Dec. 2005, pp. 10–12.

[DEO09]   I. DEMIRKOL, C. ERSOY, and E. ONUR, "Wake-Up Receivers for Wireless Sensor Networks: Benefits and Challenges," in *IEEE Wireless Communications*, vol. 16, no. 4, Aug. 2009, pp. 88–96.

[DLS+10] S. DRAGO, D. M. W. LEENAERTS, F. SEBASTIANO, L. J. BREEMS, K. A. A. MAKINWA, and B. NAUTA, "A 2.4GHz 830pJ/bit Duty-Cycled Wake-Up Receiver with -82dBm Sensitivity for Crystal-Less Wireless Sensor Nodes," in *IEEE International Solid-State Circuits Conference Digest of Technical Papers. ISSCC 2010*, Feb. 2010, pp. 224–225.

[DPF+10] M. DIELACHER, J. PRAINSACK, M. FLATSCHER, R. MATISCHEK, T. HERNDL, and W. PRIBYL, "A BAW based Transceiver used as Wake-Up Receiver," in *Proceedings of the International Conference on Architecture of Computing Systems. ARCS'10*, Hannover, Germany, Feb. 2010, pp. 235–240.

[DR11]   R. D'ERRICO and L. RUDANT, "UHF Radio Channel Characterization for Wireless Sensor Networks Within an Aircraft," in *Proceedings of the Fifth European Conference on Antennas and Propagation. EuCAP 2011*, Rome, Italy, Apr. 2011.

[DRK10]   R. D'ERRICO, L. RUDANT, and J. KEIGNART, "Channel Characterization for Intra-Vehicle WSNs in the ISM Bands," in *Proceedings of the Fourth European Conference on Antennas and Propagation. EuCAP 2010*, Barcelona, Spain, Apr. 2010, pp. 1–5.

[EEHDP04] C. C. ENZ, A. EL-HOIYDI, J.-D. DECOTIGNIE, and V. PEIRIS, "WiseNET: An Ultralow-Power Wireless Sensor Network Solution," in *Computer*, vol. 37, no. 8. IEEE Computer Society, 2004, pp. 62–70.

[EH02]   A. EL-HOIYDI, "Aloha with Preamble Sampling for Sporadic Traffic in Ad Hoc Wireless Sensor Networks," *IEEE International Conference on Communications. ICC 2002*, vol. 5, May 2002, pp. 3418–3423.

[EPH+09] R. ELFRINK, V. POP, D. HOHLFELD, T. M. KAMEL, S. MATOVA, C. DE NOOIJER, M. JANBUNATHAN, M. GOEDBLOED, L. CABALLERO, M. RENAUD, J. PENDERS, and R. VAN SCHAIJK, "First Autonomous Wireless Sensor Node Powered by a Vacuum-Packaged Piezoelectric MEMS Energy Harvester," *IEEE International Electron Devices Meeting 2009. IEDM 2009*, Baltimore, MD, Dec. 2009, pp. 1–4.

[Far08]   S. FARAHANI, *ZigBee Wireless Networks and Transceivers*. Oxford, UK: Elsevier, 2008.

[FDH+09]   M. FLATSCHER, M. DIELACHER, T. HERNDL, T. LENTSCH, R. MATISCHEK, J. PRAINSACK, W. PRIBYL, H. THEUSS, and W. WEBER, "A Robust Wireless Sensor Node for In-Tire-Pressure Monitoring," *Solid-State Circuits Conference - Digest of Technical Papers, 2009. ISSCC 2009. IEEE International*, San Francisco, CA, Feb. 2009, pp. 286–287,287a.

[FDH+10]   M. FLATSCHER, M. DIELACHER, T. HERNDL, T. LENTSCH, R. MATISCHEK, J. PRAINSACK, H. THEUSS, and W. WEBER, "A Bulk Acoustic Wave (BAW) Based Transceiver for an In-Tire-Pressure Monitoring Sensor Node," in *IEEE Journal of Solid-State Circuits*, vol. 45, no. 1, Jan. 2010, pp. 167–177.

[GAGS09]   D. GAJSKI, S. ABDI, A. GERSTLAUER, and G. SCHIRNER, *Embedded System Design: Modeling, Synthesis and Verification*. Berlin, Germany: Springer, 2009.

[GHG10]   J. GLASER, J. HAASE, and C. GRIMM, "Designing a Reconfigurable Architecture for Ultra-Low Power Wireless Sensors," *7th International Symposium on Communication Systems Networks and Digital Signal Processing. CSNDSP 2010*, Jul. 2010, pp. 311–315.

[Gut04]   J. A. GUTIERREZ, "On the Use of IEEE 802.15.4 to Enable Wireless Sensor Networks in Building Automation," *15th IEEE International Symposium on Personal, Indoor and Mobile Communications. PIMRC 2004*, vol. 3, 2004, pp. 1865–1869.

[HHJ+08]   C. HAMBECK, T. HERNDL, J. JONGSMA, F. DARRER, T. KVISTEROY, S. MAHLKNECHT, E. WESTBY, S. HUSA, E. HALVORSEN, A. VOGL, and N. P. OSTBO, "An Energy Harvesting System for In-tire TPMS," Poster presentation, *International Workshop on Power Supply On Chip*, Cork, Ireland, Sep. 2008. Available: http://www.powersoc.org/Presentations/Received/Poster P02 - Thomas Herndl - An Energy Harvesting System for In-tire TPMS.pdf

[HMH11]   C. HAMBECK, S. MAHLKNECHT, and T. HERNDL, "A 2.4 µW Wake-up Receiver for Wireless Sensor Nodes with -71 dBm Sensitivity," in *Proceedings of IEEE International Symposium on Circuits and Systems. ISCAS 2011*, Rio de Janeiro, Brazil, May 2011, pp. 534–537.

[HRW+10]   X. HUANG, S. RAMPU, X. WANG, G. DOLMANS, and H. DE GROOT, "A 2.4 GHz/915 MHz 51 µW Wake-Up Receiver with Offset and Noise Suppression," in *IEEE International Solid-State Circuits Conference Digest of Technical Papers. ISSCC 2010*, Feb. 2010, pp. 222–223.

[HSM00]   R. HEZEL, C. SCHMIGA, and A. METZ, "Next Generation of Industrial Silicon Solar Cells with Efficiencies Above 20 %," in *Conference Record of the Twenty-Eighth IEEE Photovoltaic Specialists Conference*, Sep. 2000, pp. 184–187.

[HZK+09] T. HERNDL, G. ZENNARO, J. KLAUE, P. D. BERGER, A. ALVAREZ, S. MAHLKNECHT, M. KONECNY, M. BEIGL, and W. PRIBYL, "Cooperative Hybrid Objects Sensor Networks," Poster presentation, *6th Annual IEEE Communications Society Conference on Sensors, Mesh and Ad Hoc Communications and Networks. SECON 2009*, Rome, Italy, Jun. 2009.

[JHKT09] L. JOONHYUNG, C. HANJIN, C. KOONSHIK, and P. TAHJOON, "High Sensitive RF-DC Rectifier and Ultra Low Power DC Sensing Circuit for Waking Up Wireless System," *Asia-Pacific Microwave Conference. APMC 2009*, Singapore, Dec. 2009, pp. 237–240.

[JO04] F. JONSSON and H. OLSSON, "RF Detector for On-Chip Amplitude Measurements," in *Electronic Letters*, vol. 40, no. 20, 2004, pp. 1239–1240.

[KAL05] S. KUMAR, A. ARORA, and T. H. LAI, "On the Lifetime Analysis of Always-On Wireless Sensor Network Applications," *IEEE International Conference on Mobile Adhoc and Sensor Systems Conference*, Washington, DC, Nov. 2005, pp. 186–188.

[KB06] N. P. KHAN and C. BONCELET, "PMAC: Energy Efficient Medium Access Control Protocol for Wireless Sensor Networks," *Military Communications Conference. MILCOM 2006*. IEEE, Oct. 2006, pp. 1–5.

[KB10] T. J. KAZMIERSKI and S. BEEBY, *Energy Harvesting Systems: Principles, Modeling and Applications*, 1st ed. Berlin, Germany: Springer, Nov. 2010.

[KC01] J. B. KUANG and C. T. CHUANG, "PD/SOI CMOS Schmitt Trigger Circuits with Controllable Hysteresis," in *International Symposium on VLSI Technology, Systems and Applications 2001. Proceedings of Technical Papers*, Apr. 2001, pp. 283–286.

[KL07] P. KOLINKO and L. LARSON, "Passive RF Receiver Design for Wireless Sensor Networks," *International Microwave Symposium*. IEEE, Jun. 2007, pp. 567–570.

[KPC+07] S. KIM, S. PAKZAD, D. CULLER, J. DEMMEL, G. FENVES, S. GLASER, and M. TURON, "Health Monitoring of Civil Infrastructures Using Wireless Sensor Networks," *6th International Symposium on Information Processing in Sensor Networks. IPSN 2007*, Cambridge, MA, Apr. 2007, pp. 254–263.

[KWM03] Q. A. KHAN, S. K. WADHWA, and K. MISRI, "Low Power Startup Circuits for Voltage and Current Reference with Zero Steady State Current," in *Proceedings of the 2003 International Symposium on Low Power Electronics and Design. ISLPED '03*. Seoul, Korea: IEEE, Aug. 2003, pp. 184–188.

[Lee09] E. K. F. LEE, "A Low Voltage CMOS Bandgap Reference without Using an Opamp," *International Symposium on Circuits and Systems 2009. ISCAS 2009. IEEE*, Taipei, May 2009, pp. 2533–2536.

[LEVTM09] A. LAY-EKUAKILLE, G. VENDRAMIN, A. TROTTA, and G. MAZOTTA, "Thermoelectric Generator Design Based on Power from Body Heat for Biomedical Autonomous Devices," *IEEE International Workshop on Medical Measurements and Applications. MeMeA 2009*, Cetraro, Italy, May 2009, pp. 1–4.

[LHR08] P. LE-HUY and S. ROY, "Low-Power 2.4 GHz Wake-Up Radio for Wireless Sensor Networks," *IEEE International Conference on Wireless and Mobile Computing, Networking and Communications. WIMOB 2008*, Oct. 2008, pp. 13–18.

[LHSH07] J. F. LIN, Y. T. HWANG, M. SHEU, and C. C. HO, "A Novel High-Speed and Energy Efficient 10-Transistor Full Adder Design," *IEEE Transactions on Circuits and Systems I: Regular Papers*, vol. 54, no. 5, May 2007, pp. 1050–1059.

[LLC10] J. LIM, K. LEE, and K. CHO, "Ultra Low Power RC Oscillator for System Wake-Up Using Highly Precise Auto-Calibration Technique," in *Proceedings of the ESSCIRC*, Sep. 2010, pp. 274–277.

[MAB03] G. MÖNICH, N. ANGWAFO, and G. BOECK, "A Concept for Wattless Reception in Wireless Communication between Pico Cells," *International ITG-Conference on Antennas. INICA 2003*, Berlin, Germany, Sep. 2003, pp. 21–24.

[Mah04] S. MAHLKNECHT, "Energy-Self-Sufficient Wireless Sensor Networks for Home and Buiding Environment," PhD Thesis, *Institute of Computer Technology*, Vienna University of Technology, Austria, Sep. 2004.

[MB04] S. MAHLKNECHT and M. BÖCK, "CSMA-MPS: A Minimum Preamble Sampling MAC Protocol for Low Power Wireless Sensor Networks," in *Proceedings of the IEEE International Workshop on Factory Communication Systems, 2004*, Vienna, Austria, Sep. 2004, pp. 73–80.

[MDF+10] R. MATISCHEK, M. DIELACHER, M. FLATSCHER, T. HERNDL, and J. PRAINSACK, "Optimized Protocol Processing for a Low-Power Wireless Senor Node," in *Proceedings of the International Conference on Architecture of Computing Systems. ARCS'10*, Hannover, Germany, Feb. 2010, pp. 223–228.

[MM07] S. MAHLKNECHT and S. A. MADANI, "On Architecture of Low Power Wireless Sensor Networks for Container Tracking and Monitoring Applications," in *Proceedings of the 5th International Conference on Industrial Informatics. INDIN 2007*, vol. 1, Vienna, Austria, Jun. 2007, pp. 353–358.

[MMG07] S. MADANI, S. MAHLKNECHT, and J. GLASER, "Clamp: Cross Layer Management plane for low power Wireless Sensor Networks," in *Proceedings of 5th International Workshop on Frontiers of Information Technology*, Islamabad, Pakistan, Dec. 2007.

[MRM07] C. MIN and G. A. RINCON-MORA, "Single Inductor, Multiple Input, Multiple Output (SIMIMO) Power Mixer-Charger-Supply System," *ACM/IEEE International Symposium on Low Power Electronics and Design. ISLPED 2007*, Aug. 2007, pp. 310–315.

[MSD09]   F. Maier, M. Sturmlechner, and S. Dierneder, "Novel Energy Harvester With Low Friction Losses," *6th International Multi-Conference on Systems, Signals and Devices 2009. SSD 2009*, Djerba, Tunisia, Mar. 2009, pp. 1–6.

[NT02]   W. Nosovic and T. D. Todd, "Scheduled Rendezvous and RFID Wakeup in Embedded Wireless Networks," *IEEE International Conference on Communications. ICC 2002*, vol. 5, May 2002, pp. 3325–3329.

[OCB+10]   C. Oestges, N. Czink, B. Bandemer, P. Castiglione, F. Kaltenberger, and A. J. Paulraj, "Experimental Characterization and Modeling of Outdoor-to-Indoor and Indoor-to-Indoor Distributed Channels," in *IEEE Transactions on Vehicular Technology*, vol. 59, no. 5, Jun. 2010, pp. 2253–2265.

[OCL+04]   B. P. Otis, Y. H. Chee, R. Lu, N. M. Pletcher, and J. M. Rabaey, "An Ultra-Low Power MEMS-Based Two-Channel Transceiver for Wireless Sensor Network," in *Symposium on VLSI Circuits 2004. Digest of Technical Papers*, Honolulu, USA, Jun. 2004, pp. 20–23.

[OCR05]   B. Otis, Y. H. Chee, and J. Rabaey, "A 400 µW-RX, 1.6 mW-TX Super-Regenerative Transceiver for Wireless Sensor Networks," *Solid-State Circuits Conference - Digest of Technical Papers, 2005. ISSCC 2005. IEEE International*, San Francisco, CA, Feb. 2005, pp. 6–7.

[PGR07]   N. Pletcher, S. Gambini, and J. M. Rabaey, "A 65 µW, 1.9 GHz RF to Digital Baseband Wakeup Receiver for Wireless Sensor Nodes," in *Proc. Custom Integrated Circuits Conference. CICC 2007*. IEEE, Sep. 2007, pp. 539–542.

[Ple08]   N. M. Pletcher, "Ultra-Low Power Wake-Up Receivers for Wireless Sensor Networks," PhD Thesis, *University of California*, Berkeley, CA, 2008.

[PRG09]   N. Pletcher, J. Rabaey, and S. Gambini, "A 52 µW Wake-Up Receiver With -72 dBm Sensitivity Using an Uncertain-IF Architecture," in *IEEE Journal of Solid-State Circuits. JSSC 2009*, vol. 44, no. 1, Jan. 2009, pp. 269–280.

[RA05]   C. Rossi and P. Aguirre, "Ultra-low Power CMOS Cells for Temperature Sensors," *18th Symposium on Integrated Circuits and Systems Design 2005*, Florianopolis, Sep. 2005, pp. 202–206.

[Rab09]   J. Rabaey, "The Standby Power Challenge: Wake-up Receivers to the Rescue," *International Symposium on VLSI Technology, Systems, and Applications. VLSI-TSA '09*. IEEE, Apr. 2009, p. 42.

[RM04]   K. Römer and F. Mattern, "The Design Space of Wireless Sensor Networks," in *IEEE Wireless Communications*, vol. 11, no. 6, Dec. 2004, pp. 54–61.

[RSK+07]   J. Rybicki, B. Scheuermann, W. Kiess, C. Lochert, P. Fallahi, and M. Mauve, "Challenge: Peers on Wheels - A Road to New Traffic Information Systems," in *Proceedings of the 13th Annual International Conference on Mobile*

Computing and Networking. MobiCom 2007, Montreal, Canada, Sep. 2007, pp. 215–221.

[SBB08] T. SHIHABUDHEEN, V. S. BABU, and M. R. BAIJU, "A Low Power Sub 1V 3.5-ppm/°C Voltage Reference Featuring Subthreshold MOSFETs," *15th IEEE International Conference on Electronics, Circuits and Systems 2008. ICECS 2008*, St. Julien's, Aug. 2008, pp. 442–445.

[SBK+06] M. SHEETS, F. BURGHARDT, T. KARALAR, J. AMMER, Y. H. CHEE, and J. RABAEY, "A Power-Managed Protocol Processor for Wireless Sensor Networks," in *Symposium on VLSI Circuits 2006. Digest of Technical Papers*, 2006, pp. 212–213.

[SD09] M. SPINOLA-DURANTE, "Wakeup Receiver for Wireless Sensor Networks," PhD Thesis, *Institute of Computer Technology*, Vienna University of Technology, Austria, Dec. 2009.

[SDM09] M. SPINOLA-DURANTE and S. MAHLKNECHT, "An Ultra Low Power Wakeup Receiver for Wireless Sensor Nodes," *Third International Conference on Sensor Technologies and Applications. SENSORCOMM '09*, Athens, Greece, Jun. 2009, pp. 167–170.

[SKP00] B. SAHOO, M. KUHLMANN, and K. K. PARHI, "A Low-Power Correlator," in *Proceedings of the 10th Great Lakes Symposium on VLSI. GLSVLSI '00*, 2000.

[SMZ07] K. SOHRABY, D. MINOLY, and T. ZNATI, *Wireless Sensor Networks: Technology, Protocols and Applications*. Hoboken, New Jersey: John Wiley & Sons, 2007.

[STGS02] C. SCHURGERS, V. TSIATSIS, S. GANERIWAL, and M. SRIVASTAVA, "Optimizing Sensor Networks in the Energy-Latency-Density Design Space," *IEEE Transactions on Mobile Computing*, vol. 1, no. 1, 2002, pp. 70–80.

[SWP10] R. SU, T. WATTEYNE, and K. S. J. PISTER, "Comparison between Preamble Sampling and Wake-Up Receivers in Wireless Sensor Networks," *Conference on IEEE Global Telecommunications. GLOBECOM 2010*, Miami, FL, Dec. 2010, pp. 1–5.

[TSM+09] T. TAKIGUCHI, S. SARUWATARI, T. MORITO, S. ISHIDA, M. MINAMI, and H. MORIKAWA, "A Novel Wireless Wake-Up Mechanism for Energy-efficient Ubiquitous Networks," *IEEE International Conference on Communications Workshops. ICC Workshops 2009*, Jun. 2009, pp. 1–5.

[vdMB07] S. VON DER MARK and G. BOECK, "Ultra Low Power Wakeup Detector for Sensor Networks," *Microwave and Optoelectronics Conference 2007. IMOC 2007. SBMO/IEEE MTT-S International*, Brazil, Dec. 2007, pp. 865–868.

[vdMKHB05] S. VON DER MARK, R. KAMP, M. HUBER, and G. BÖCK, "Three Stage Wakeup Scheme for Sensor Networks," *IEEE International Conference on Microwave and Optoelectronics 2005*, Jul. 2005, pp. 205–208.

[VGGC07] X. VALLVE, A. GRAILLOT, S. GUAL, and H. COLIN, "Micro storage and Demand Side Management in distributed PV grid-connected installations," *9th International Conference on Electrical Power Quality and Utilisation. EPQU 2007*, 2007, pp. 1–6.

[Vit03] E. A. VITTOZ, "Weak Inversion in Analog and Digital Circuits," Presentation, *CCCD Workshop 2003*. Lund, Sweden: Competence Center for Circuit Design, Oct. 2003.

[Vit09] ———, "Weak Inversion for Ultra Low-Power and Very Low-Voltage Circuits," *Solid-State Circuits Conference. A-SSCC 2009*. Taipei, Taiwan: IEEE Asia, Nov. 2009, pp. 129–132.

[VWS+06] G. VIRONE, A. WOOD, L. SELAVO, Q. CAO, L. FANG, T. DOAN, Z. HE, R. STOLERU, S. LIN, and J. A. STANKOVIC, "An Assisted Living Oriented Information System Based on a Residential Wireless Sensor Network," in *Proceedings of the 1st Distributed Diagnosis and Home Healthcare Conference*, Arlington, Virginia, Apr. 2006, pp. 95–100.

[XJSCSG08] J. XIAOHUA, L. JEONG-SEON, S. CHANG, and L. SANG-GUG, "A 53 µW Super-Regenerative Receiver for 2.4 GHz Wake-Up Application," *Asia-Pacific Microwave Conference. APMC 2008*, Hong Kong, Dec. 2008, pp. 1–4.

[XLHK11] W. XU, Y. LI, Z. HONG, and D. KILLAT, "A 90 % Peak Efficiency Single-Inductor Dual-Output Buck-Boost Converter with Extended-PWM Control," in *IEEE International Solid-State Circuits Conference Digest of Technical Papers. ISSCC 2011*, San Francisco, CA, Feb. 2011, pp. 394–396.

[ZSA03] C. ZHANG, A. SRIVASTAVA, and P. K. AJMERA, "Low voltage CMOS Schmitt trigger circuits," in *IET Electronics Letters*, vol. 39, no. 24, Nov. 2003, pp. 1696–1698.

Internet References

[1] CHOSeN Research Project. (2011) Homepage. http://www.chosen.eu

[2] e-Cubes Research Project. (2011) Homepage. http://ecubes.epfl.ch

[3] Intitute of Computer Technology, TU Wien. (2009) PAWiS Research Project Homepage. http://www.ict.tuwien.ac.at/pawis

[4] Infineon Technologies AG. (2011) Homepage. http://www.infineon.com

[5] European Aeronautic Defence and Space Company EADS. (2011) Homepage. http://www.eads.com

[6] CHOSeN Project report. (2011, Jun.) Description of the Automotive Demonstrator. http://www.chosen.eu/userfiles/Chosen_D1 3a_Description of the automotive demonstrator(PU)_v1_37.pdf

[7] Intitute of Computer Technology, TU Wien. (2011, May) WCMS Research Project Homepage. http://www.ict.tuwien.ac.at/wcms

[8] Atmel Corporation. (2011) Homepage. http://www.atmel.com

[9] Texas Instruments Incorporated. (2011) Homepage. http://www.ti.com

[10] ZigBee Alliance, Inc. (2011) Homepage. http://www.zigbee.org

[11] Microchip Technology Incorporated. (2011) Homepage. http://www.microchip.com

[12] STMicroelectronics. (2011) Homepage. http://www.stm.com

[13] Analog Devices Incorporated. (2011) Homepage. http://www.analog.com

[14] NXP Semiconductors. (2011) Homepage. http://www.nxp.com

[15] ARM Holdings, plc. (2011) Homepage. http://www.arm.org

[16] Philips Technologie GmbH, U-L-M Photonics. (2011) Homepage. http://www.ulm-photonics.com

[17] Avago Technologies. (2011, Jun.) PCB Layout Guidelines for Designing with Avago SFP+ Transceivers, Application Note 5362. http://www.avagotech.com/docs/AV02-0725EN

[18] Linx Technologies Inc. (2011) Homepage. http://www.linxtechnologies.com

[19] Linear Technology. (2011) Homepage. http://www.linear.com

[20] Nexergy. (2010, Dec.) Battery Chemistry Comparison Chart. http://www.nexergy.com/media/pdfs/batterychemchart.pdf

[21] Cymbet Corporation. (2011) Homepage. http://www.cymbet.com

[22] Infinite Power Solutions Inc. (2011) Homepage. http://www.infinitepowersolutions.com

[23] Saft S.A. (2011) Homepage. http://www.saftbatteries.com

[24] Maxwell Technologies Inc. (2011) Homepage. http://www.maxwell.com

[25] Murata Manufacturing Co. Ltd. (2011) Homepage. http://www.murata.com

[26] Renata SA. (2010, Dec.) General Technical Information on Lithium Batteries. http://www.renata.com/content/encapsulated/general.pdf

[27] Schott solar. (2011, Jul.) EFG-Solar Cell. http://www.schott.com/photovoltaic/english/download/efg_125x125_e_0706.pdf

[28] Linz Center of Mechatronics GmbH. (2011) Homepage. http://www.lcm.at

[29] IMEC Belgium. (2011) Homepage. http://www.imec.be

[30] Micropelt GmbH. (2011) Homepage. http://www.micropelt.com

[31] Torex Semiconductor Ltd. (2011) Homepage. http://www.torex-europe.com

[32] Micro Crystal AG. (2011) OV-7604-C7 Datasheet. http://www.microcrystal.com

[33] Euroquartz Ltd. (2011) Homepage. http://www.euroquartz.co.uk

[34] Internet Engineering Task Force. (2011, Jul.) 6LoWPAN documentation. http://datatracker.ietf.org/wg/6lowpan

[35] Austria Microsystems AG. (2011) Homepage. http://www.austriamicrosystems.com

[36] Cadence Design Systems, Inc. (2011) Homepage. http://www.cadence.com

[37] The MathWorks, Inc. (2011) Homepage. http://www.mathworks.com

List of Abbreviations

μC	microcontroller
ADC	analog-to-digital converter
AES	advanced encryption standard
ASIC	application specific integrated circuit
ASK	amplitude shift keying
BAW	bulk acoustic wave
BB	baseband
BER	bit error ratio
BGR	bandgap reference
BiCMOS	bipolar and complementary metal-oxide semiconductor
BOD	brownout detection
BPF	bandpass filter
BPSK	binary phase shift keying
CCP	capture/compare/PWM
CDMA	code division multiple access
CHOSeN	Cooperative Hybrid Objects Sensor Networks
CMOS	complementary metal-oxide semiconductor
CPU	central processing unit
CRC	cyclic redundancy check
CSMA/CA	carrier sense multiple access/collision avoidance
DAC	digital-to-analog converter
DC	direct current
DCO	digitally controlled oscillator
DMA	direct memory access
DRAM	dynamic random access memory
EMU	energy management unit
ESD	electrostatic discharge
ESR	equivalent series resistance
FEA	frontend amplifier
FPGA	field programmable gate array
FSK	frequency shift keying
FWR	false wake-up rate
GPIB	general purpose interface bus

GPIO	general purpose input-output
GPRS	general packet radio service
GPS	global positioning system
GSM	global system for mobile communications
I	in-phase component
I^{2}C	inter-integrated circuit
I^{2}S	integrated interchip sound
IC	integrated circuit
IF	intermediate frequency
IR	image rejection
IrDA	Infrared Data Association
IRQ	interrupt request
ISM	industrial scientific and medical
ISO	International Organization for Standardization
ITU-R	International Telecommunication Union Radiocommunication Sector
LDO	low drop output voltage regulator
LED	light emitting diode
LF	low frequency
LNA	low noise amplifier
LO	local oscillator
LSB	least significant bit
MAC	medium access control
MCU	micro controller unit
MEMS	micro electromechanical system
MIPS	million instructions per second
MLCC	multi-layer chip capacitor
MOS	metal-oxide semiconductor
MPU	memory protection unit
MSB	most significant bit
MUX	multiplexer
NMOS	n-channel metal-oxide semiconductor
NVIC	nested vector interrupt controller
OFDM	orthogonal frequency division mulitplex
OOK	on-off keying
OSI	open systems interconnection
PA	power amplifier
PAM	pulse amplitude modulation
PAWiS	Power Aware Wireless Sensors
PC	personal computer
PCB	printed circuit board
PGA	programmable gain amplifier
PLA	programmable logic array
PLL	phased locked loop
PMOS	p-channel metal-oxide semiconductor
PPM	pulse position modulation

List of Abbreviations

PSRR	power supply rejection ratio
PTAT	proportional to absolute temperature
PWM	pulse width modulation
Q	quadrature component
QoS	quality of service
RAM	random access memory
RC	resistor-capacitor
REF	reference
RF	radio frequency
RFID	radio frequency identification
RISC	reduced instruction set computer
RMS	root-mean-square
RSSI	received signal strength indication
RTC	realtime clock
RX	receiver
SAR	successive approximation register
SAW	surface acoustic wave
SC	switched-capacitor
SD	secure digital
SiP	system-in-package
SNR	signal-to-noise ratio
SoC	system-on-chip
SPI	serial seripheral interface
SSI	synchronous serial interface
TC	temperature coefficient
TDMA	time division multiple access
TEM	thermoelectric module
TPMS	tire pressure monitoring system
TWI	two-wire interface
TX	transmitter
UART	universal asynchronous receiver transmitter
UMTS	universal mobile telecommunications system
USB	universal serial bus
USN	ubiquitous sensor network
WDT	watchdog timer
WLAN	wireless local area network
WSN	wireless sensor network
WuR	wake-up receiver
XTAL	crystal

Die VDM Verlagsservicegesellschaft sucht für wissenschaftliche Verlage abgeschlossene und herausragende

Dissertationen, Habilitationen, Diplomarbeiten, Master Theses, Magisterarbeiten usw.

für die kostenlose Publikation als Fachbuch.

Sie verfügen über eine Arbeit, die hohen inhaltlichen und formalen Ansprüchen genügt, und haben Interesse an einer honorarvergüteten Publikation?

Dann senden Sie bitte erste Informationen über sich und Ihre Arbeit per Email an *info@vdm-vsg.de*.

Sie erhalten kurzfristig unser Feedback!

VDM Verlagsservicegesellschaft mbH
Dudweiler Landstr. 99
D - 66123 Saarbrücken
www.vdm-vsg.de

Telefon +49 681 3720 174
Fax +49 681 3720 1749

Die VDM Verlagsservicegesellschaft mbH vertritt

Printed by Books on Demand GmbH, Norderstedt / Germany